[英国]蒂莫西·高尔斯 著 刘熙 译

牛津通识读本·

数学

Mathematics

A Very Short Introduction

译林出版社

图书在版编目（CIP）数据

数学 /（英）高尔斯（Gowers, T.）著；刘熙译. —南京：译林出版社，2014.3
(2024.12重印)
（牛津通识读本）
书名原文：Mathematics: A Very Short Introduction
ISBN 978-7-5447-4523-9

Ⅰ.①数… Ⅱ.①高…②刘… Ⅲ.①数学－研究 Ⅳ.①O1-0

中国版本图书馆 CIP 数据核字（2013）第 242823 号

Mathematics: A Very Short Introduction, by Timothy Gowers
Copyright © Timothy Gowers, 2002
Mathematics: A Very Short Introduction was originally published in English in 2002. This licensed edition is published by arrangement with Oxford University Press. Yilin Press, Ltd is solely responsible for this bilingual edition from the original work and Oxford University Press shall have no liability for any errors, omissions or inaccuracies or ambiguities in such bilingual edition or for any losses caused by reliance thereon.
Chinese and English edition copyright © 2014 by Yilin Press, Ltd
All rights reserved.

著作权合同登记号　图字：10-2014-197 号

数学　[英国] 蒂莫西·高尔斯　/ 著　刘　熙　/ 译

责任编辑　许　丹　何本国
责任印制　董　虎

原文出版　Oxford University Press, 2002
出版发行　译林出版社
地　　址　南京市湖南路 1 号 A 楼
邮　　箱　yilin@yilin.com
网　　址　www.yilin.com
市场热线　025-86633278
排　　版　南京展望文化发展有限公司
印　　刷　南通印刷总厂有限公司
开　　本　890 毫米 × 1260 毫米　1/32
印　　张　9.375
插　　页　4
版　　次　2014 年 3 月第 1 版
印　　次　2024 年 12 月第 21 次印刷
书　　号　ISBN 978-7-5447-4523-9
定　　价　39.00 元

版权所有 · 侵权必究

译林版图书若有印装错误可向出版社调换　质量热线：025-83658316

序言
李大潜

数学是绝大多数人学得最多的一门功课,但对于"数学是什么?"这一看来很普通的问题,却很难一下子给出一个使公众满意的回答。按照恩格斯的说法,数学是以现实世界的空间形式和数量关系为研究对象的。尽管人们现在对空间形式和数量关系的理解已经大大深化和拓展了,但作为一种哲学的概括,恩格斯的这一论断应该说并没有过时,也便于向公众表述并为公众所接受。然而,要真正深入地回答"数学是什么?"这个问题,不能仅仅从定义出发,而必须涉及数学的具体内涵,作一些比较深入的解释和说明,才能达到使人信服的效果。但是,要这样做,会常常碰到下面两个似乎难以克服的技术上的困难。

第一,数学内涵的展现离不开众多的术语、记号和公式。在公众对有关的数学内涵产生兴趣并开始有所领悟之前,很可能早已为这些术语、记号和公式搞得晕头转向甚至望而却步了。

第二,数学内涵的展现同样离不开必要的逻辑推理。推理若过分严密,很难引起公众的兴趣;但若过于粗疏,语焉不详,则又易使人不得要领。

在现在的这一本书中,看不到很多的术语、记号和公式,对有关的数学概念及内涵一律用简明而生动的语言来介绍,看似如数家珍,娓娓道来,但举重若轻,高屋建瓴,反而更好地揭示了本质。不熟悉有关数学内容的读者,会感到茅塞顿开、豁然开朗;而已经熟悉有关内容的读者,也会有如沐春风、别开生面的感受。书中在论述极限时,用有限来刻画无限的生动而细致的处理方式,虽然本质上是经典的 $\varepsilon-N$(或 $\varepsilon-\delta$)数学描述的一套直观而通俗的说法,但脱离了数学记号和公式,更显得清楚和亲切,就是一个典型的范例。书中对黄金分割比是一个无理数的证明、用地图册来介绍微分流形的概念等等,都同样独具匠心,可圈可点。另一方面,作者在不便于或不需要进行严格数学推导的时候,会巧妙地绕过去,但对于必要的推导和论证,绝不偷工减料,而是不遗余力地以十分详尽的方式加以说明,一步步地将读者引向有关的结论,同时也加深了读者对逻辑论证严密性的理解。在这方面,关于数系的扩张、关于 $\sqrt{2}$ 是无理数的证明、关于平行公设的独立性等段落,都是倾心着力撰写的。正因为作者作了如此认真的努力,这一本篇幅不大的书显得出类拔萃,应该说为现有众多的数学科普读物提供了一个楷模。能够做到这一点,而且做得如此出色,不言而喻,归根结底是和作者深厚的数学功力、对数学内涵及其精神实质的深刻理解和把握分不开的。

本书一开始就讲了数学建模,指出数学研究的对象只是有关现实世界的数学模型,同时,指出有关的数学模型并不真正是相应的现实世界,而只是它的一个近似的代表与反映。在书中作者反复强调并解释的是他的一个基本的观点:对于数学,不要问它是什么,而只要问它能做什么。这一抽象化的思考方法,将重点放在数学内部体系的相容性,强调新的数学概念、方法与内容和已有的数学体系应自然地融为一体，强调要将有关的数学内容脱离其物理上的实在、变为符合一些特定规则的记号，就会更利于应用，更利于正确地理解高等的数学。作者在书中举出指数函数、圆的面积、高维空间、分数维的几何等等一系列的例子来阐明这一观点对克服难关，深入理解与拓展数学概念所带来的好处。这是很有启发性的，也是很自然的，反映了抽象思维的必要性和优越性。由于有关的数学模型虽然只是现实世界的一个近似的代表，但毕竟是一个代表，适应于它的一些规则一开始并不是凭空而来的，而是从现实世界中移植、挪用或抽象过来的，对不同的现实世界可能引入不同的规则，也会造成不同的数学对象（不满足乘法交换律的矩阵，就是一个例子）。数学与现实世界的关系，套用一句文艺界的术语，看来应该是源于生活、高于生活的关系。如果作者在强调他的上述观点及做法的同时，也能够强调，数学要真正得到原创性的重大进展，除了要密切关注及面对数学内部的矛盾运动外，还要密切关注现实世界（包括其他科学技术）对数学提出的问题和需求，努力从外部世界中汲取生动活泼、丰富多彩的营养，应努力使二者相互促进、相得益彰，是不是会更全面、更富有启发性呢？

本书的作者蒂莫西·高尔斯（Timothy Gowers）教授是 1998 年获得菲尔茨（Fields）奖的著名数学家。2000 年当我在法国巴黎访问时，因美国克雷（Clay）研究所给法国科学院院士阿兰·科纳（Alain Conne）教授颁发一个大奖，曾在法兰西学院（Collège de France）的讲演大厅里召开过颁奖会。会上获菲尔茨奖不久的蒂莫西·高尔斯教授应邀作了一个公众讲演。他在强调数学是一个整体的时候，曾说，如果把所有的数学分支按是否有联系组成一个网络，一定是一个连通的网络，而不会有一些学科，尽管它们看来与其他分支联系很少，游离于整个数学这一大网络之外。这正像有些人有很多亲戚朋友，有些则很少，但整个社会的人群所组成的网络仍是连通的一样。他的这一观点及如此通俗易懂的说法曾给我留下了深刻的印象，从这个意义上说，我和他已有一面之缘。这次有机会看到他这一本颇具特色的数学科普著作的中译本问世，也是一件幸事，特为之序。

2013 年 12 月 25 日

目录

前言 1

1 模型 1

2 数与抽象 16

3 证明 35

4 极限与无穷 56

5 维度 70

6 几何 85

7 估计与近似 109

8 常见问题 122

索引 134

英文原文 139

前言

20 世纪初,伟大的数学家大卫·希尔伯特发现,有很多数学中的重要论点在结构上十分类似。他意识到,在适当的广义范畴下,这些论点事实上可以视为等同。与此类似的一系列发现为一个崭新的数学分支开启了大门。而这一新领域中的一个核心概念——希尔伯特空间——正是以希尔伯特的名字来命名的,这一概念使许许多多的现代数学研究变得清晰,范围之广包括了从数论直到量子力学各个分支,以至于如果你对希尔伯特空间的基本理论一无所知,你就根本不能算是一名受过良好教育的数学家。

那么,什么是希尔伯特空间呢? 在典型的高校数学课程中,它被定义为"完备的内积空间"。修读这样一门课程的学生,理应从先修课程中了解到,所谓"内积空间"是指配备了内积的向量空间,而所谓"完备"是指空间中任意柯西列都收敛。当然,要想理解这样的定义,学生还必须知道"向量空间"、"内积"、"柯西列"和"收敛"的定义。就拿其中一个举例来说(这还并不是最长的一个):序列 $x_1, x_2, x_3, \cdots$ 若满足对于任意正数 ϵ,总存在整数 N,使得对于任意大于 N 的整数 p 和 q,x_p 与 x_q 间的距离不大于 ϵ,则称这个序列为柯西列。

简言之，如果你希望了解希尔伯特空间是什么，你就必须首先学习并且消化一系列由低到高、等级分明的较低级概念。毫无疑问这需要耗费时间和精力。对于许多最重要的数学思想来说都是这样。有鉴于此，要写一本书提供对数学的简单易懂的介绍，其所能达到的目标就极为有限，更何况这本书还需要写得很短。

我没有选择用更聪明的办法绕着这个难题走，而是集中关注数学交流中另一重完全不同的障碍。这重障碍并非技术性的，而更多属于哲学性质的。它区分开了两种人：一种人乐于接受诸如无穷大、负一的平方根、第二十六维和弯曲空间这样的概念，另一类人则觉得这些概念荒诞不经。其实无须沉浸在技术细节中，依然有可能坦然接受这些思想，我将努力表明如何做到这一点。

如果说这本书要向你传达什么信息的话，那就是——我们应当学习抽象地思考，因为通过抽象地思考，许多哲学上的困难就能轻易地消除。在第二章里，我将详细说明什么是抽象的方法。第一章中则考虑我们更熟悉、与日常更相关的抽象：从现实世界的问题中提取核心特征，从而将其转化为数学问题的过程。第三章中我将讨论什么叫作"严格的证明"。这前三章是关于一般性的数学的。

之后我将讨论一些更加具体的课题。最后一章与其说是关于数学的，不如说是关于数学家的，因此会跟前几章有些不同。我建议你在读过第二章后再阅读后续章节。除此以外，这本书已经尽量做到不受先后顺序影响——在任何章节中，我并没有假定读者已经理解并记住了先前的内容。

读这本书并不需要太多的预备知识，英国 GCSE 课程①或同等水平即可。不过我假定读者具有一些兴趣，而不是需要靠我去大力宣扬。因此，我在书中没有用到趣闻轶事、漫画、惊叹号、搞笑的章节标题或者曼德布罗特集合②的图片。我同样避免了混沌理论、哥德尔定理等内容：与它们在当前数学研究中的实际影响相比，这些内容在公众的想象中所占的比例已经过大，而且其他图书已经充分地阐释了这些内容。我所选择的内容都是很普通的，详细地去讨论，以说明怎样通过一种更深刻的方式来理解它们。换言之，我的目标在深不在广，在于向读者传达主流数学的魅力，让读者体会到它的不言而喻。

感谢克雷数学研究所和普林斯顿大学在我写作此书期间对我的支持和热情接待。感谢吉尔伯特·阿代尔、丽贝卡·高尔斯、埃米莉·高尔斯、帕特里克·高尔斯、乔书亚·卡茨和埃德蒙·托马斯阅读了本书的初稿。他们非常聪明，知识丰富，实在不能算作普通读者，不过还是能够让我放心，至少某些非数学专家是能够读懂我的作品的。基于他们对此书的评论，我作出了许多改进。我把这本书献给埃米莉，希望她能够借此了解一点点我整天都在做的是些什么事情。

---

① 约相当于初中水平。——译注，下同

② 分形的一种。

第一章
模型

扔石头问题

风轻云淡的一天,你站在水平地面上,手里拿着一块石头,想要扔得越远越好。已知你能用多大的力气扔出去,那么最重要的决策就是选择石头出手时与地面的夹角。如果夹角太小,那么尽管石头在水平方向的速度分量很大,也会很快落到地面上,因而飞不出太远;反之,若夹角过大,石头能在空中停留较久但掠过的水平距离却不远。很明显我们需要在这中间作一些权衡。

利用牛顿物理学和微积分的一些初步知识,可以计算得到最佳的折中方案——石头离手时应与地面呈 45 度夹角。就这个问题而言,这基本上是最简洁优美的答案了。同样的计算还可以告诉我们石头在空中的飞行轨迹是个抛物线,甚至还能得出脱手后在空中任意时刻的速度有多大。

看起来,科学与数学相结合能够使我们预测石块飞出去直至落地之前的一切行为。然而,只有在我们作了许多的简化假设之后才能够如此。其中最主要的假设是,作用在石头上的只有一种力,即地球的引力,而且这种力的大小及方向在各处总是一样的。但实际上并非如此,因为它忽略了空气阻力、地球自转,也没有计入月球的微弱引力,而且越到高处地球引力越小,在地球表面上"垂直向下"的方向也随着具体位置的不同而逐渐变化。即使你能够接受上述计算,45度角的结果也基于另一个隐含假设:石头离手的初始速度与夹角无关。这也是不正确的:实际上夹角越小,人能使上的力气越大。

上述这些缺陷的重要性各有不同,我们在计算和预测中应该采取怎样的态度来对待这些偏差呢?把所有因素全部考虑在内进行计算固然是一种办法,但还有一种远为明智的办法:首先决定你需要达到什么样的精确度,然后用尽可能简单的办法达到它。如果经验表明一项简化的假设只会对结果产生微不足道的影响,那就应当采取这样的假设。

例如,空气阻力的影响相对来说是比较小的,因为石头很小很硬,密度大。假如在出手角度上有较大的误差,那么通过计入空气阻力来将计算复杂化就没有多大意义。如果一定要考虑进去的话,以下这条经验法则就足矣:空气阻力变大,则通过减小出手角度来弥补。

何为数学模型

当我们考察一个物理问题的解答时,十有八九能够就其中科学贡献部分和数学贡献部分划出一道清晰的界线。科学家在观察和实验的基础上,作一些简洁性与解释有效性的一般性考虑,建立一种理论。数学家,或者做数学的科学家,则研究理论的纯粹逻辑结果。有时候,这些情形是常规计算的结果,常规计算所预言的现象正是理论在提出时所要解释的。在某些偶然的情况下,理论所作出的预言则完全出乎意料。如果这些意料之外的现象后来被实验所证实,那么我们就得到了支持这种理论的重要证据。

然而,由于我上面所讨论到的简化问题,"证实一项科学预言"的概念就多多少少有了些问题。让我们考虑另一个例子:牛顿的运动定律和引力定律告诉我们,两个物体从同样的高度开始作自由落体运动,它们将同时到达地面(如果地面平坦)。这种现象由伽利略首先提出,它有点违背我们的直觉。实际上,它违背的不仅是我们的直觉:如果你亲自试一试,比方说用高尔夫球和乒乓球,你会发现高尔夫球首先落地。既然如此,究竟在什么意义上伽利略的论断是正确的呢?

当然,由于空气阻力的存在,我们不可能把这个小实验当作伽利略理论的反例:实验证明,当空气阻力很小时理论是正确的。如果你对此有所怀疑,觉得空气阻力实在稀松平常,怎能总是挽救牛顿力学的预测于失败之际,那么,找个机会去观察一下羽毛在真空中的下落,你就能重拾对科学的信念以及对伽利略的赞赏——真空中,羽毛的下落的确与石头别无二致。

尽管如此,由于科学观察永远不是完全直接性和决定性的,我们仍需要一种更好的方式来描述科学与数学之间的关系。数学家并不是将科学理论直接应用于现实世界中,而是应用于**模型**上。在这里,模型可以看作是所要研究的那部分现实世界的一种虚构、简化的版本。在模型里,我们就有可能进行完全精确的计算。在扔石头的例子中,现实世界与模型的关系正如同图1和图2所展示的那样。

对于一种给定的物理情形,有多种方法将其模型化。我们需要结合切近的经验与深入的理论考量来决定,哪种模型更有可能向我们透露世界的本真。选择模型时,有一个需要优先考虑的因素,即模型的行为应当与实际中观察到的行为密切对应。

图 1　飞行中的球甲

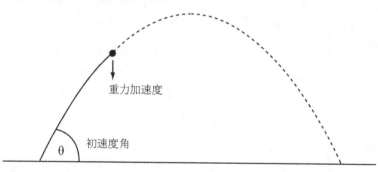

图 2　飞行中的球乙

但是，诸如简洁、数学表达上的优雅等其他因素可能反而时常会更重要一些。实际上，确实有一些模型在现实世界中几乎找不到任何相似之物，但同时却非常有用。接下来有一些例子将会对此进行说明。

掷骰子问题

假设我要掷一对骰子,想要了解它们的行为如何。经验告诉我,问某些问题根本是不现实的。例如,不可能期待有人能预先告诉我某一次掷骰子的结果,即便是他掌握了很高超的科技,并且用机器来掷骰子。与此相反的是,关于概率的问题则常常是能够回答的,比如"两个骰子的结果之和为 7 的可能性有多大"之类的问题。这样的问题的答案可能也是有用处的,比方说在我玩双陆棋[①]赌钱的时候。这一类问题很容易模型化,只要将两个骰子掷出来的结果看作是从下列 36 个整数对中随机选取一个。

(1,1) (1,2) (1,3) (1,4) (1,5) (1,6)
(2,1) (2,2) (2,3) (2,4) (2,5) (2,6)
(3,1) (3,2) (3,3) (3,4) (3,5) (3,6)
(4,1) (4,2) (4,3) (4,4) (4,5) (4,6)
(5,1) (5,2) (5,3) (5,4) (5,5) (5,6)
(6,1) (6,2) (6,3) (6,4) (6,5) (6,6)

每组数对中第一个数表示骰子甲的结果,第二个数表示骰子乙的结果。恰有六组满足两数之和为 7,因此掷出 7 的概率就是 6/36,即 1/6。

可能有人会反对这种模型,他们会说,骰子在滚动时是遵循牛顿定律的,至少在很高的精度上遵循,因此骰子落地的情况根本不是随机的:原则上是完全能够被计算出来的。但是,"原则上"这个短语在这里被过度使用了,因为这样的计算将会是

---

① 一种棋子游戏,靠掷骰子决定行棋格数。

极端复杂的,并且需要知道骰子的形状、材料、初始速度、旋转速度等更为精确的信息,而这般精确的信息在实际中是根本无法测出来的。因为这一点,使用某种更为复杂的决定论模型是无论如何也不会有任何优势的。

预测人口增长

较"软"性的科学——比如生物学和经济学中,也充斥着各种数学模型,这些模型都远比它们所要真正表示的现象简单得多,甚至以某些方式刻意地使其不够精确,但是这些模型还是有其用场、富于启发性。就以一个在经济学上有重要意义的生物学问题为例,我们来考虑预测一国未来20年的人口。我们可能会用到一种非常简单的模型,即将全国人口表示为一组数对$(t, p(t))$,其中t表示时间,$p(t)$表示时刻t的人口规模。另外我们要用到两个数b和d,来表示出生率和死亡率。所谓出生率和死亡率,即每年出生人数和死亡人数占总人口的比例。

假设我们已知2002年初的总人口是p。根据上述模型,2002年的出生人数和死亡人数将分别为bp和dp,因此2003年初的总人口将为$p+bp-dp=(1+b-d)p$。其他年份亦然,因此我们就能够写出公式$p(n+1)=(1+b-d)p(n)$,意即$n+1$年年初的人口是n年年初人口乘以$(1+b-d)$。换句话说,每一年人口数量都会乘上$(1+b-d)$。那么20年后的人口就是乘以$(1+b-d)^{20}$,于是就得出了初始问题的答案。

这个模型已经比较好了,它能向我们证明,如果出生率明显高于死亡率,那么人口就会急剧增长。但即便如此,它也还是不够现实的,它作出的预测可能很不精确。比方说,模型中假设出生率和死亡率在20年中都保持不变,这并不太可信。过去的事实已经证明，出生率和死亡率经常会受到社会变迁和政治事件的影响，如医学进步、新型疾病出现、女性首次生育年龄增大、税负激励以及偶尔发生的大规模战争等等。生育率与死亡率会随时间变化还有另一个原因，就是一国国民的年龄分布可能相当失衡。比方说，15 年前出现了一波婴儿潮，那么我们就有理由预期再过 10 年到 15 年出生率就会增加。

因此，通过引入其他因素来使模型复杂化，这个想法相当诱人。我们记出生率和死亡率分别为 $b(t)$ 和 $d(t)$，使其可以随时间变化。我们并不想单用一个数字 $p(t)$ 来表示总人口，我们可能想要知道不同年龄层各有多少人。如果同时还能尽可能多地知道各个年龄层的社会态度和行为倾向，也会对预测未来的出生率和死亡率有所帮助。获取这样的统计信息是十分昂贵且困难的，但这些信息确实能够大幅提高预测的精度。因此，没有一种模型能够脱颖而出，声称比其他模型都好。关于社会和政治的变迁，谁也不可能确切地说出情况会是什么样子。关于某种模型，我们所能提出的合理问题，最多只能是问某种有条件的预测，也就是说模型只能告诉我们，这样的社会或政治变迁如果发生的话会产生怎样的影响。

气体的行为

气体动理论由丹尼尔·伯努利在 1738 年提出，后来又由麦克斯韦、玻尔兹曼等人在 19 世纪后半叶推进。根据这种理论，气体是由运动着的分子组成的，气体的许多性质——如温度和压强，都是这些分子的统计属性。譬如，温度就对应着分子的平均速度[①]。

---

① 原文如此，有误，应为"平均动能"。

有了这样的想法之后，让我们设想一种模型来描述方盒子中的气体。这个盒子当然应该用一个立方体来表示(意即数学的而非物理的)。既然分子是非常小的，那么用立方体中的点来表示也就很自然了。这些点应当是运动的，所以我们必须确定控制它们运动的规则。此时，我们需要作出一些选择。

如果盒中只有一个分子，那么规则可以很明显：分子以恒定速度运动，撞到盒子壁面时就反弹出去。要将这种模型推广到包含 N 个分子的情形(N 是个较大的数)，最简单的办法就是假设分子都遵从这样的运动规则，分子之间绝对没有相互作用。为了启动这样的 N 分子模型，我们要选择分子(或者说，表示它们的那些点)的初始位置及初始速度。随机选择是一种好办法，因为我们可以预期，在任意时刻，真实气体中的分子都在空间中弥散着，运动方向也各式各样。

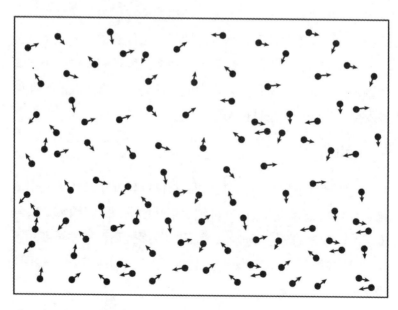

图 3　气体的二维模型

要说清在立方体中随机取一个点并不困难，随机的运动方向也不复杂，但如何随机地选择速率就有些含混了，因为速率可以取从 0 到无穷大的任意值。为了避免这个困难，我们可以作一个从物理角度看似不太可信的假设，让所有分子的速度大小都相同，仅仅让初始位置和方向能够随机选取。图 3 就表示了这个模型的一个二维情形。

N 个分子完全相互独立运动的假设毫无疑问是过度简化的。比方说，利用这个模型，我们就不可能理解，为什么当温度足够低时气体会液化：当你把模型中的各点运动速率降低，得到的还是相同的模型，无非跑得慢一些而已。不过这个模型还是能够解释真实气体的许多行为。例如，想象盒子被慢慢压缩的情形。分子仍然会继续以相同速率运动，但由于盒子变小，分子撞击壁面更加频繁，可供撞击的壁面面积也变小了。由于这两个缘故，单位面积的壁面每秒钟被撞击次数就增多了。这些撞击正是气体压强的来源，于是我们可以总结出，气体体积减小时，气体压强很可能增大——正如实际观测所证实的那样。类似的论证还可以解释，为什么气体温度升高而体积不变时，压强会增大。要推算出压强、温度与体积之间的数值关系也并不困难。

上述模型大致上就是伯努利所提出的模型。麦克斯韦的成就之一就是发现了一个优美的理论，来解决如何更逼真地选择初始速率的问题。为了理解这一点，让我们放弃分子间没有相互作用的假设。作为替代，我们假定分子会时不时地相互碰撞，就像台球一样。碰撞之后，它们就以另外的速率、向另外的方向，在遵守能量守恒和动量守恒定律的前提下随机弹开。当然，既然我们用没有体积的点来表示分子，那么就很难看出它们要如何碰撞。不过，这个麻烦在理论中恰可以作为一个非正式的论据，说明分子运动速度及方向具有某种随机性。麦克斯韦就这种随机性的本质作了两个非常合理的假定:其一,分子运动的随机性不随时间而改变;其二,这个随机性在不同方向上没有区别。大体来讲,第二条假设意味着,选取 d_1 和 d_2 两个方向及某个速率 s,那么粒子以速率 s 沿着 d_1 方向运动的概率和以速率 s 沿着 d_2 方向运动的概率是相同的。不可思议的是,这样的两个假设就足以恰好决定分子运动速度的分布形式,即意味着,如果我们想要随机选取速度,就只有一种自然的方式。(它们应当服从正态分布。这种分布产生了著名的"钟形曲线"。这种曲线在各种各样的场合下经常出现,既出现在数学中也出现在实验中。)

一旦选定了速度,我们就可以再次忘掉分子间的相互作用。结果表明,这种作了一点改进的模型依然存在着原始模型中的许多瑕疵。为了进一步修正,我们只能再把分子间的相互作用考虑进来。但是结果发现,即使是非常简单的相互作用粒子模型,其行为也极其复杂,会引发极为难解,事实上多数都未能解决的数学问题。

大脑和计算机的模型化

计算机同样也可以看作由相互作用的各部分集合而成;很大程度上由于这个原因,理论计算机科学中同样有很多悬而未决的重要问题。其中有如下这样一个例子,我们可能愿意去尝试解答。假设某人选取了两个素数 p 和 q,将它们相乘后的结果 pq 告诉你。你只要逐个取素数,看看它是否能够整除 pq,即可以找出 p 和 q。例如,给出 91,你很快就能发现它不是 2,3,5 的倍数,继而发现它恰好等于 7×13。

然而,当 p 和 q 非常大时——比方说都是 200 位的素数,那么这一试错过程会耗时极长，即使借助于强力计算机也是如此。(如果你想要体会一下这种困难，不妨尝试找出 6901 的两个素因子，以及 280 123 的。)可另一方面，似乎也不难感觉到，说不定这个问题存在着更聪明的解决办法，基于它就可以编制出一种快速运转的计算机程序。如果能找到这种好办法，我们就能破解作为大部分现代安全系统之基石的密码，包括在互联网上以及其他各处——破解这些密码的难点就在大整数的因子分解。反之，如果能够表明由 pq 计算出 p 和 q 的这种快速有效的方法不存在的话，我们则能够安心。不幸的是，虽然计算机总在不断地让我们惊叹它的各种能力，对于它们做不了的，我们却几乎毫无了解。

在思考这个问题之前，我们必须找到一种方法来数学化地表示计算机，并且要尽可能简单。图 4 所显示的就是一种极好的方法。它包含许多层节点，连结点和点之间的线段称作"边"。进入到最顶层的称作"输入"，这是一条 0 和 1 的序列，从最底层出来的叫作"输出"，是另一条 0 和 1 的序列。节点分为三种，分别称作"与门"、"或门"和"非门"。每一个门都从连结上层的边中接收到一些 0 和 1。它再根据所接收到的数码来自己发出一些 0 和 1，所遵循的简单规则如下：与门当接收到的输入全部为 1 时，输出 1，否则输出 0；或门当接收到的输入全部为 0 时，输出 0，否则输出 1；非门只允许一条边连结上层，它在接收到 1 时输出 0，接收到 0 时输出 1。

一系列门由边连接起来就称作一条**电路**，我上面所描述的模型正是关于计算的电路模型。使用"计算"一词是恰如其分的，因为我们可以把电路看作这样一种装置，它拥有一条 0 和 1 的序列，继而按某些预定规则将其变换为另一条序列，如果电

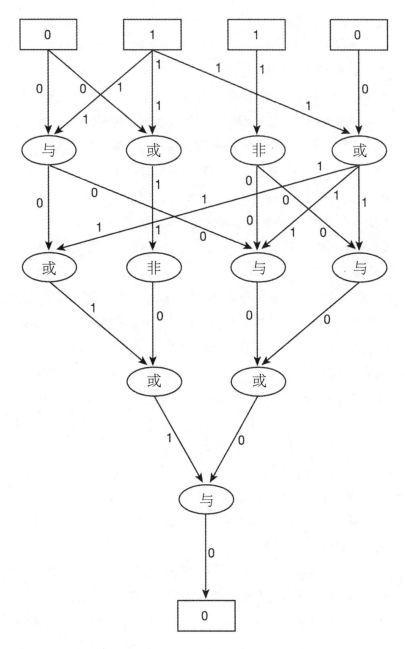

图4 一个简单的计算机程序

路很大，变换规则可能会很复杂。这也正是计算机所进行的工作，只不过它们能够把这些序列翻译成我们能够理解的格式，诸如高级程序设计语言、视窗系统、图标等等。实际上，存在一种比较简单的办法（仅从理论的角度而言——在实践上操作将是个噩梦），能将任意计算机程序转换为一条按完全相同的规则变换 01 序列的电路。而且，计算机程序的重要特征恰在其对应电路中有着非常类似的对照物。

具体而言，电路中的节点数量正对应于计算机程序运转所花费的时间。因此，如果我们能够表明，按某种方式来变换 01 序列需要庞大的电路，那么也就说明这种变换方式所需要的计算机程序运转时间很长。我们使用电路模型而非直接分析计算机，其优势就在于，从数学的角度来看电路更简单也更自然，考虑起来更容易一些。

对电路模型进行一点小小的修改，我们就能得到大脑的一种有用的模型。这种模型不再使用 01 序列，而是使用 0 和 1 之间的任意值来表示强度各异的信号。所有的门，即对应于神经元或者脑细胞，也有所不同，但其行为还是很简单的。每个门从其他的门接收到一些信号，如果这些信号的总强度——对应数字的总和——足够大，门就在某个特定的强度水平上输出它自己的信号，否则不输出。这对应于神经元所作的是否"激发"的决策。

似乎很难相信这个模型能够捕捉大脑全部的复杂性，但这部分缘于我并没有提到应当有多少个门以及如何安排这些门。一个典型的人类大脑包含大约 1000 亿个神经元，它们以非常复杂的方式排列着；以我们当前对大脑的认识，还不可能谈及太多——至少在精微的细节方面不可能说清。不过，上述模型提供了一种有益的理论框架，供我们思考大脑可能是如何工作的,也使我们能够模拟某些类似于大脑运行的行为。

地图染色与时间表制定

设想你正在绘制一幅地图,地图上分成了若干区域,你希望为这些区域选取颜色。你可能想选用尽可能少的颜色,但同时还希望避免任意两块相邻区域使用相同的颜色。再设想你正在安排大学课程的时间表。课程有很多门,但可供安排的总时间段有限,所以会有某些门课程时间冲突。哪些学生选了哪些课程已经登记在列,你希望尽可能合理安排,仅当两门课程没有学生同时选择时才可以时间冲突。

这两个问题看似截然不同,但一种合理的模型能够说明,从数学的观点来看它们其实是一样的。在这两个问题中,都需要给一些对象(国家、课程)赋予一些属性(颜色、时间)。对象中有某些两两组合(相邻的国家,不能冲突的课程)是不能相容的,也就是说它们不能被赋予相同的属性。在这两个问题中,我们其实并不关心具体的对象是什么、要赋予的属性是什么,所以我们也可

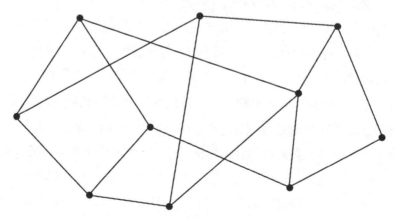

图5　10个顶点和15条边组成的图

以仅用点来表示它们。为了表示那些不相容的成对的点，我们可以将它们用线段连结起来。这样一组边和边连结起来的点的集合，就是"图"这种数学结构。图5给出了一个简单的例子。通常称图中的点为顶点，称线段为边。

一旦我们将问题用这种形式表示出来，我们在两个例子中的任务就统一为：将顶点分成尽可能少的几组，使得每组中不包含由同一条边相连的两顶点。（图5所表示的图可以分成三组，但不可能分成两组。）这也就说明了使模型尽可能简化的另一个充足理由：如果幸运的话，同样的模型可以用来一次性研究很多不同的现象。

"抽象"一词的不同含义

在设计模型的时候，我们会忽视所考察的现象中尽可能多的信息，从中仅仅抽象出那些对理解其行为必不可少的特征。在上述我所讨论的例子中，石头被简化为一个点，一国的全部人口被简化为一个数，大脑被简化为遵循一些简单数学规则的门的网络，分子间相互作用被简化到根本不存在。结果得到的数学结构就是具体情形在模型化之后的抽象表示。

我们说数学是一个抽象的领域，这包含两层含义：一来它从问题中抽象出重要特征，二来它所处理的对象不是具体的、有形的。在下一章，我们将讨论数学抽象的第三层，也是更深层的含义，前面的例子其实已经让我们对此有所了解。图是一种具有可塑性的模型，可以用在多种场合。但当我们研究图时，完全不需要考虑它的这些具体用途：点究竟表示地区、课程还是别的完全不同的东西，这并不重要。研究图的理论工作者可以完全抛弃现实世界，进入到纯粹抽象的王国之中。

第二章
数与抽象

抽象方法

几年前,《泰晤士报文学增刊》的一篇评论在开篇写道:

> 已知 $0\times 0=0$ 以及 $1\times 1=1$,就可以得到:平方等于自身的数是存在的。但是再进一步,我们就可以得到:数是存在的。经过这简简单单朴实无华的一步,我们似乎就从一个基本的算术式得到了一个令人吃惊的、极具争议的哲学结论:数是存在的。你可能还以为这有多么困难呢。
>
> A.W. 莫尔评论杰罗尔德·J.卡茨《现实理性主义》,
> 《泰晤士报文学增刊》,1998 年 9 月 11 日

对于这个论证,我们可以从许多角度来批评,而且基本上不可能有人把它当真,包括评论人自己。然而,确实有哲学家在严肃地思考数是否存在这样的问题,而且正是这种思考使得他们不同于数学家——数学家理所当然地认为数就是存在的,不理解它何以能够成为一个问题。本章的主要目的就是要解释,为什么数学家能够愉快地忽视这样一个看似非常基本的问题,其

至理应忽视它。

如果我们把同样的论证方法搬到国际象棋上来,这个"简简单单朴实无华"的关于数的存在的论证就明显荒谬了。已知象棋中的黑色国王有时可以斜向移动一格,由此得出,存在着可以斜向移动一格的棋子。但接下来将进而得出:国际象棋的棋子是存在的。当然了,我想说的并不是那句平平常常的断言:人们有时候会制造国际象棋的棋具。毕竟,没有棋具同样也能下棋。我所指的是一个更加"惊人"的哲学结论:象棋棋子的存在独立于它们的物理形态。

在象棋中,黑色国王是什么呢?这是个奇怪的问题。最令人满意的处理方式似乎是,回避这个问题。除了指着棋盘,说明游戏的规则,我们还能做些什么呢?也许在这么做的同时会对黑色国王给予特别的关注?真正重要的问题并不是黑色国王的存在性或者它的本质属性,而是它在游戏中所发挥的作用。

人们所谓的数学中的抽象方法,正是我们采取类似态度来对待数学对象的结果。这种态度能够用这样一句话涵盖:数学对象**是**其所**做**。在语言哲学中,我们能发现类似的话经常出现,而且可能饱受争议。我们可以举出两个例子,比如索绪尔所说的"语言中只有差异",还有维特根斯坦所说的"语词的意义是它在语言中的用法"(参见书末"延伸阅读")。另外,我们还可以加上逻辑实证主义者的宣言:"陈述的意义就是其证实的方法。"如果你出于哲学方面的考虑,认为我的观点令人生厌,那你不妨不要将它看作一个教条式的断言,而是视为一种可供采纳的态度。实际上,我希望表明的是,要想正确地理解更高等的数学,采纳这种态度是至关重要的。

没有棋子的象棋

看出这一点很有意思——尽管我的论述并不直接依赖于它:象棋,或者任何类似的游戏,都可以以图为模型。(图已经在前一章末定义过了。)图的顶点代表游戏的某种可能的局面。如果两个顶点 P 和 Q 有边相连,那就意味着可以从局面 P 出发,经过合乎规则的一步之后达到局面 Q。因为有可能无法从 Q 返回到 P,所以这样的边需要用箭头来指示方向。某些顶点可以看作白棋获胜,还有某些顶点可以看作黑棋获胜。游戏从一个特定顶点,即游戏的开始局面出发。两位棋手相继沿着边移动。第一位棋手努力要走到白棋获胜的某个顶点,第二位棋手则要走到黑棋获胜的顶点。图 6 显示了某种大大简化的类似游戏。(图中不难看到,对于这个游戏来说白棋有必胜策略。)

尽管象棋的这种图论模型很难有应用上的意义,因为现实中可能的局面数量实在太庞大了,但是就其所表示的游戏和象棋完全等价来说,它仍然是一种完美的模型。不过在定义它的时候,我完全没有提及关于棋子的任何事情。从这个角度来看,

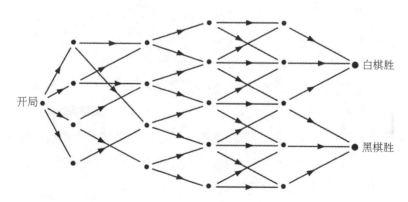

图 6 *白先,白方有必胜策略*

黑色国王是否存在的问题就是个离奇的问题了：棋盘和棋子只不过是方便我们将这么大的图中一团乱麻似的顶点和边组织起来的一种原则而已。如果我们说出"黑色国王被将军了"这样的句子，那么这只是一种简化的说法，它所指的无非是两位棋手到达了极多的顶点中的某一个。

自然数

"自然"是数学家对我们所熟悉的 1, 2, 3, 4 这样的数字所赋予的称呼。自然数是最基本的数学对象，但它们似乎并没有引导我们去抽象地思考。毕竟，单单一个数字 5 能**做**些什么呢？它不可能像个棋子一样走来走去。它所具有的似乎是一种内在的属性——某种纯粹的"五性"，当我们观察图 7 这样的图片时就能立即从中提炼出这样的性质。

然而，当我们考虑更大的数时，其中的纯粹性就变少了。图 8 向我们表示了 7, 12 和 47 这几个数。可能有部分人能够立刻把握第一张图中的"七性"，但大多数人可能会在一瞬间有这样

图 7 "五性"的概念

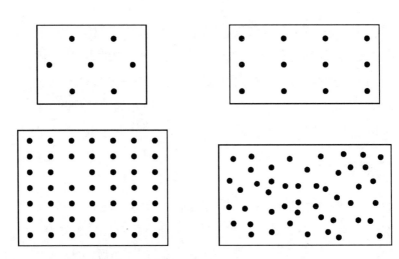

图8 表示7,12和47(两种)的方式

的思考:"外围的点形成了一个六边形,再加上中心的一个得到6+1=7。"类似地,12可能会被考虑为3×4或2×6。至于47,和别的数字比起来,比方和46相比,一组这个数量的物体就缺乏特别之处。如果这组物体以某种模式排列起来,例如排成少两个点的7×7的阵列,那么我们的知识就可以通过7×7-2=49-2=47迅速地得出总数一共有多少。如若不然,我们就只能去一个个地数,这时我们则是将47视为46之后的那个数,而46又是45之后的那个数,依此类推。

也就是说,当数字变得还不算太大时,我们就已经不再将其视作一些独立的客体了,而开始通过它们的内在属性,它们与其他数字的关联,以及它们在**数系**中的作用来理解。这也就是我之前说数能"做"什么所要表达的意思。

如我们已经清楚地看到的,数的概念与加法、乘法这样的算术运算紧密相连。比方说,如果没有算术的概念,对1 000 000 017这样的数的意义就只能有很模糊的把握。一个数系并不仅仅是一堆数字,而是由数字及算术规则共同构成的。我们还可以这样来总结这种抽象方法:考虑规则,而不是考虑数字本身。按这种观点,数字就可以被当作某种游戏中的记号(或许应该被称为计数子)。

为了对其中的规则有一些了解,让我们来考察一个简单的算术问题:如果我们要确证 $38\times263=9994$,应当做什么?多数人可能会用计算器来检查,但要是因故无法这样检查呢?他们可能会作如下的推理。

$$38\times263=30\times263+8\times263$$
$$=30\times200+30\times60+30\times3+8\times200+8\times60+8\times3$$
$$=6000+1800+90+1600+480+24$$
$$=9400+570+24$$
$$=9994$$

为什么上述几步看起来是如此地天经地义呢?比方说,为什么我们会不假思索地相信 $30\times200=6000$? 30 的定义是 3×10,200 的定义是 $2\times(10\times10)$,所以我们可以充分相信 $30\times200=(3\times10)\times(2\times(10\times10))$。但为什么是 6000 呢?

一般没人会去问这种问题,不过既然有人问了,我们可能会说:

$$(3\times10)\times(2\times(10\times10))=(3\times2)\times(10\times10\times10)=6\times1000=6000$$

我们并没有经过仔细考虑就利用了关于乘法的两个熟知事

实:其一,两数相乘时谁先谁后并没有关系;其二,多个数相乘时无论怎样加括号都没有区别。例如,7×8=8×7 以及(31×34)×35=31×(34×35)。第二个例子中,中间结果肯定会受到括号位置的影响,但我们知道最终结果是相同的。

这两条规则被称为乘法的**交换律**和**结合律**。下面我将列出若干条加法和乘法中常用的规则,也包括上述两条。

A1 加法交换律:对任意两个数 a 和 b,有 $a+b=b+a$。

A2 加法结合律:对任意三个数 a、b 和 c,有 $a+(b+c)=(a+b)+c$。

M1 乘法交换律:对任意两个数 a 和 b,有 $ab=ba$。

M2 乘法结合律:对任意三个数 a、b 和 c,有 $a(bc)=(ab)c$。

M3 1 是乘法单位元:对任意数 a,有 $1a=a$。

D 分配律:对任意三个数 a、b 和 c,有 $(a+b)c=ac+bc$。

我列出这些规则并非试图告诉你这些规则本身多么有意思,而是想请你关注它们在我们的思考中所扮演的角色——即便是在相当简单的数学陈述中。我们对 $3×2=6$ 的信心大概是基于这样的一种图像:

$$\begin{array}{ccc} * & * & * \\ * & * & * \end{array}$$

这样一来,要直接论证 $38×263=9994$ 就是不可能的了,于是我们要以完全不同的方式来思考这一稍显复杂的事实,其中就要利用到交换律、结合律和分配律。如果的确遵守了这些规则,我们就会相信最后的结果。而且,虽然绝不可能对 9994 个物体有视觉上的感知,我们也相信结果是正确的。

零

在历史上,数字 0 的思想的诞生晚于正整数。这在很多人看来是个神秘而且矛盾的概念,它引发了类似这样的问题:"某种事物既然没有,怎么可能存在呢?"但从抽象的观点来看,0 其实很明确:它只不过是引入到我们数系中的一个新记号而已,并且满足下面这条特殊的性质。

A3 0 是加法单位元:对任意数 a,有 $0+a=a$。

这就是关于 0 你所需要知道的一切了。无关它意味着什么,只是一条小规则告诉你它做什么。

那关于数字 0 的其他性质呢,比方说 0 乘以任何数都等于 0 的性质?我并没有列出这条规则,因为我们可以从性质 **A3** 及之前的其他规则把它推导出来。例如,我们已经将 2 定义为 1+1,那要怎样来表明 $0 \times 2=0$ 呢?首先,根据规则 **M1** 有 $0 \times 2=2 \times 0$。然后,由规则 **D** 得到 $(1+1) \times 0=1 \times 0+1 \times 0$。但根据规则 **M3**,$1 \times 0=0$,所以该式等于 $0+0$。规则 **A3** 意味着 $0+0=0$,于是我们的论证就完成了。

从非抽象的角度出发,可能会这样去论证:0×2 的意思是指,**总共 0 个 2 相加**,没有 2,就是 0。但用这种思考方式,我们将不太容易回答诸如我儿子约翰问我的这个问题(在他六岁时):既然无和无相乘的意思是**没有无**,那为什么结果又会是无呢?尽管当时可能不太适合,但一个好的回答终归是,它能够如下所述从基本规则中推导出来。(每一步之后我都列出了所用到的规则。)

$0 = 1 \times 0$	**M3**
$= (0+1) \times 0$	**A3**
$= 0 \times 0 + 1 \times 0$	**D**
$= 0 \times 0 + 0$	**M3**
$= 0 + 0 \times 0$	**A1**
$= 0 \times 0$	**A3**

为什么我要对非常基本的事实给出如此冗长的证明呢？和上面一样，原因并不是我觉得这些证明多么有数学趣味，而是想表明，抽象地(利用几条简单规则，忽略数字的具体意义)而非具体地(考察数学陈述的实际意义)证明算术陈述是怎么一回事。将实际意义及思维图像与数学对象结合起来固然非常有用，但是，正如我们将多次在本书中看到的，这样的结合常常并不足以告诉我们在新的不熟悉的场合下应当怎样去处理。因而，抽象的方法是不可或缺的。

负数和分数

谁如果曾经给小孩子教过数学，那一定会知道，减法和除法并不像加法和乘法那样直接，他们学习起来更困难一些。为了解释减法，我们当然可以利用"拿走"的概念，提这一类的问题："一开始有 5 个橘子，吃掉 2 个，还剩下几个？"然而这并不总是思考减法的最佳方式。例如，当我们用 100 减 98 时，更好的想法不是从 100 中取走 98，而是考虑什么数加上 98 能够得到 100。

也就是说，更有效的做法是求解方程 98+x=100，尽管计算时字母 x 并不常出现在我们的脑子里。类似地，考虑除法也有两种方式。为了解释 50 除以 10 的意义，我们既可以问"50 个东西分成相等的 10 组，每组几个"，也可以问"10 个东西分一组，50 个东西能分几组"。第二种办法等于在问"10 和几相乘能够得到 50"，也就等于解方程 10x=50。

向小孩解释减法和除法还有着更进一步的困难，那就是这两种计算并非总能够进行。例如，不可能从装有 7 只橘子的碗中拿走 10 只，11 颗弹珠不可能平均分给 3 个小孩。但成人却能够计算 7 减去 10 和 11 除以 3，分别得到 −3 和 11/3。问题随之而来：−3 和 11/3 这样的数实际上存在吗？如果存在它们又是什么呢？

从抽象的角度看，处理这个问题的方式类似于之前对于零的处理方式：全都抛至脑后。关于 −3，我们只需要知道它和 3 相加等于 0 即可；关于 11/3，只需知道它乘以 3 等于 11 即可。这就是它们的运算规则。再与之前的规则相结合，我们就可以在更大的数系中进行算术运算。为什么我们希望照这样扩充数系呢？原因就是，这样得到的模型允许我们在其中求解 $x+a=b$ 和 $ax=b$ 这样的方程，无论 a 和 b 取何值——只要 a 在第二个方程中不为 0。换句话说，这样得到的模型中，减法和除法总是能够进行的，只要除数不为 0。（除数为 0 的问题我们会在本章稍后谈到。）

按照这样的路数，我们只需再增加两条规则来扩充我们的数系：一条给我们带来负数，另一条给我们带来分数，即我们所熟知的**有理数**。

A4 加法逆元：对任意数 a，总存在一个数 b 使得 $a+b=0$。

M4 乘法逆元：对任意不为 0 的数 a，总存在一个数 c 使得 $ac=1$。

确定了这些规则后，我们就可以将 $-a$ 和 $1/a$ 分别看作 **A4** 中的 b、**M4** 中的 c 的代号。至于更一般的表达式 p/q，它表示 p 乘以 $1/q$。

规则 **A4** 和 **M4** 还蕴含了另外两条规则，即消去法则。

A5 加法消去律：对任意三个数 a、b 和 c，若 $a+b=a+c$，则 $b=c$。

M5 乘法消去律：对任意三个数 a、b 和 c，若 $ab=ac$ 且 a 不为 0，则 $b=c$。

第一条可以通过在等式两端加上 $-a$ 得到证明，第二条可以通过在等式两端乘以 $1/a$ 得到证明。应当注意 **A5** 和 **M5** 的地位与之前的那些规则是不同的——这两条是之前规则的**推论**，而不是我们向游戏中引入的新规则。

如果有人要加两个分数，如 2/5 加 3/7，那么常用的方法是找出它们的公分母：

$$\frac{2}{5}+\frac{3}{7}=\frac{14}{35}+\frac{15}{35}=\frac{29}{35}$$

这种方法及类似方法的合理性可以用我们的新规则来证明。例如：

$$35 \times \frac{14}{35} = 35 \times \left(14 \times \frac{1}{35}\right) = (14 \times 35) \times \frac{1}{35} = 14 \times \left(35 \times \frac{1}{35}\right)$$
$$= 14 \times 1 = 14$$

又有

$$35 \times \frac{2}{5} = (5 \times 7) \times \left(2 \times \frac{1}{5}\right) = (7 \times 5) \times \left(\frac{1}{5} \times 2\right)$$
$$= \left(7 \times \left(5 \times \frac{1}{5}\right)\right) \times 2 = (7 \times 1) \times 2 = 7 \times 2 = 14$$

于是，由规则 **M5** 得到 2/5 和 14/35 是相等的，正如我们在计算中所假设的那样。

类似地，我们还可以论证关于负数的一些熟悉的事实。请读者自己练习从上述规则中推出 $(-1) \times (-1) = 1$，它的推导和对 $0 \times 0 = 0$ 的证明相当类似。

为什么在很多人看来，负数的实在性要低于正数呢？大概因为对数量不多的物体的计数是人类的基本活动，在这其中并不会用到负数。但这只不过意味着，作为模型的自然数系在某种特定场合下比较有用，而扩充数系则不太用得上。但如果我们考虑温度、日期或者银行账户，那负数就**的确**能够发挥作用了。只要扩充数系是逻辑自洽的——实际上也正是如此，用它作为模型就没有任何害处。

把自然数系称作一种模型似乎有点奇怪。难道我们不是在**切实地数数**，未涉及任何特定的理想化描述吗？我们的确是那样数数的,但数数的过程并非总是恰当的,甚至会根本不可能。从数学的观点来看,1 394 840 275 936 498 649 234 987 这个数没有任何问题,但如果我们连佛罗里达州的选票都数不过来,就无法想象能确信自己拥有由 1 394 840 275 936 498 649 234 987 个东西组成的一堆。如果你将两堆落叶加到第三堆上,得到的结果并不是三堆树叶,而是一大堆树叶。倘若你刚观察过一场暴雨,那正如维特根斯坦所说,"'你看到了多少水滴'这个问题的恰当答案是,**很多**。并不是因为没有那么一个数字,而是因为你根本不知道有多少"。

实数和复数

实数系包含了所有能够用十进制无穷小数表示的数字。这个概念看似简单,实则不易,其中的缘由我们会在第四章加以解释。而现在,让我们来讨论一下将有理数系扩充到实数系的原因。我要讲的是,这原因正与引入负数和分数的理由类似:它们使我们能够求解某些方程,缺少它们我们则无法求解。

这些方程中最著名的一个莫过于 $x^2=2$。在公元前 6 世纪,毕达哥拉斯学派发现 $\sqrt{2}$ 是无理数,即它不能表示为一个分数。(我们将在下一章中给出证明。)当时这项发现激起了一片错愕,但时至今日我们已经能够欣然接受:要想将正方形对角线的长度之类的事物模型化,就必须扩充我们的数系。抽象方法再一次使我们的任务变得轻易。我们引入一种新的符号——$\sqrt{2}$,并引入一条说明它能够做什么的新规则:它的平方等于 2。

如果你对此颇有研究,你可能会反对我刚才说的,理由是,这样的规则并不能区分$\sqrt{2}$和$-\sqrt{2}$。处理这个问题的办法之一就是向我们的数系中引入一个新的概念——**序**。比较数和数之间的大小总是有用处的,而且这还能使我们通过$\sqrt{2}$的额外的性质——大于0——来识别它。不过即便没有这种性质,我们也已经能够做一些运算,例如:

$$\frac{1}{\sqrt{2}-1} = \frac{\sqrt{2}+1}{(\sqrt{2}-1)(\sqrt{2}+1)} = \frac{\sqrt{2}+1}{(\sqrt{2})^2-\sqrt{2}+\sqrt{2}-1}$$

$$= \frac{\sqrt{2}+1}{2-1} = \sqrt{2}+1$$

而且,**不区分**$\sqrt{2}$和$-\sqrt{2}$其实是有一点好处的——上面的计算对这两个数都是成立的。

从用来描述数系每次扩充所得到的新数的名字,我们能够发现在历史上对抽象方法质疑的一些痕迹,比如"负的"和"无理的"。但更让人难以下咽的还在后面,这就是"虚幻的",或者"复杂的"数,即形如$a+bi$的数,其中a和b均为实数,i是-1的平方根。

从具象的观点来看,我们会很快就摈弃-1的平方根:因为任何数的平方都是正的,-1根本就没有平方根,故事到此为止。然而,若采纳了抽象的观点,这种反对意见就显得软弱无力了。只要引入方程$x^2=-1$的解并把它称作i就好了,为什么不继续单纯地把数系扩充下去呢?为什么偏偏它的引入就应该比之前的$\sqrt{2}$更值得反对呢?

一种回答大概是,$\sqrt{2}$能够按十进制小数展开,(原则上)

能够计算到任意精度,而 i 就与此不同了。但这说的只不过是我们已经知道的事情,即 i 不是实数——正如 $\sqrt{2}$ 不是有理数一样。这并不能阻挡我们扩充数系,在其中进行如下的运算:

$$\frac{1}{i-1} = \frac{i+1}{(i-1)(i+1)} = \frac{i+1}{i^2-i+i-1} = \frac{i+1}{-1-1} = -\frac{1}{2}(i+1)$$

i 和 $\sqrt{2}$ 之间最主要的区别就是我们**被迫**抽象地去思考 i,而对于 $\sqrt{2}$ 我们则还有备选方案,可以将它具体地表示为 $1.4142\cdots$,或者看作单位正方形的对角线长度。要看出为什么 i 没有这样的表示方法,不妨问问自己这个问题:-1 的两个平方根中,哪个是 i 哪个是 $-i$ 呢? 这个问题是没有意义的,因为我们对 i 所定义的**唯一**的性质就是平方等于 -1。既然 $-i$ 也有同样的性质,那么关于 i 成立的那些命题,如替换为关于 $-i$ 的相应命题,必定依然成立。一旦领会了这一点,就很难再赞同 i 指示一个独立存在的实在的客体。

这和一个著名的哲学难题有相似之处。你对红色所产生的感受与我对绿色产生的感受(交换亦可)有没有可能是相同的呢? 一些哲学家很严肃地思考这个问题,并定义"感受性"一词来表示我们所拥有的绝对的内在体验,比如我们对色彩的体验。而另一些人并不相信感受性。在他们看来,"绿色"这样的词有更抽象的定义,那就是根据它在语言系统中所发挥的作用,也就是说,根据它与"草地"、"红色"等概念之间的关系。因此,就这个论题,要想从人们谈论色彩的方式来推断出他们的态度是不可能的,除非在哲学争论当中。类似地,在实践中,关于数和其他数学对象,重要的只是它们所遵循的规则。

如果说为了使方程 $x^2=-1$ 有解我们引入 i，那么其他类似的方程呢？比如 $x^4=-3$ 或者 $2x^6+3x+17=0$ 呢？值得注意的是，人们发现，所有这样的方程都可以在复数系中求解。也就是说，我们通过接受 i 作出小小的投资，结果得到了许多倍的回报。发现这个事实的历史过程有点复杂，但人们通常将它归功于高斯。这个事实被人们称为代数基本定理，它给我们提供了令人折服的证据，使我们相信 i 的确有合情合理、自然而然的地方。我们的确无法想象一个篮子里有 i 只苹果，车行途中经过了 i 个小时，银行账户透支了 i 英镑。但对数学家来说，复数系已经必不可少。对科学家和工程师同样也是。比如，量子力学的理论就高度依赖于复数。复数作为最佳的例证之一，向我们表明了一条概括性原则：一种抽象的数学构造若是充分自然的，则基本上必能作为模型找到它的用途。

初探无穷大

一旦我们学会抽象地思考，事情就会立刻变得令人愉悦，这个境况有点像突然能够骑自行车而不必去担心保持平衡。然而，我也并不希望使读者觉得抽象方法就好像是印钞许可证。我们可以来作一个有趣的对比，比较一下向数系中引入 i 与引入数字无穷大之间的区别。乍看起来，似乎没什么可以阻止我们的：无穷大应当用来表示 1 除以 0 之类的，所以，为什么不使 ∞ 成为一个抽象符号，用它来表示方程 $0x=1$ 的解呢？

但当我们想做算术时，这个想法的问题立刻就来了。我们在这里举个例子，利用乘法结合律 **M2** 和 $0 \times 2=0$ 的事实，就可以得到简单推论：

$$1 = \infty \times 0 = \infty \times (0 \times 2) = (\infty \times 0) \times 2 = 1 \times 2 = 2$$

这个式子表明，方程 $0x=1$ 的解若存在将会导致**不相容性**。这是否意味着无穷大不存在呢？并不是。这只说明，无穷大的自然概念与算术定律是不相容的。扩充数系以将符号 ∞ 包含进来，并且接受在新的系统下这些算术定律并非总是成立，这样做有时是有用处的。但是通常，人们还是希望保持算术定律，不考虑无穷大。

把负数和分数放到指数上

抽象方法还有一点极大的优越性：它使我们能够将熟悉的概念扩展到不熟悉的情况下，赋予其新的意义。我所说的"赋予意义"的确是恰当的，因为我们所做的正是去赋予意义，而不是去发现某种早就存在的意义。这当中有一个简单的例子，那就是我们如何扩展指数的概念。

如果 n 是个正整数，那么 a^n 即表示 n 个 a 相乘的结果。如 $5^3 = 5 \times 5 \times 5 = 125$ 以及 $2^5 = 2 \times 2 \times 2 \times 2 \times 2 = 32$。但若以此作为定义，我们就不容易去解释 $2^{3/2}$ 这样的表达式，因为你不可能拿出一个半的 2，把它们乘在一起。那处理这种问题的抽象方法是什么呢？我们又一次地需要抛开寻找内在意义的意识。在本例中即需要忽视 a^n 的内在意义，转而考虑关于它的规则。

关于指数运算的两条基本规则是：

E1 对任意实数 a，$a^1 = a$。

E2 对任意实数 a 和任意一对自然数 m、n，有 $a^{m+n} = a^m \times a^n$。

例如，$2^5=2^3\times2^2$，因为 2^5 表示 $2\times2\times2\times2\times2$，而 $2^3\times2^2$ 表示 $(2\times2\times2)\times(2\times2)$，由乘法结合律可知两数相同。

从上述两条规则出发，我们可以迅速重新得到已经知道的一些事实。比如，根据 **E2** 即知 $a^2=a^{1+1}$ 等于 $a^1\times a^1$；再根据 **E1**，此即为 $a\times a$，正如我们所了解的。除此以外，我们现在还能够做更多的事情。让我们用 x 来表示 $2^{3/2}$。那么 $x\times x=2^{3/2}\times2^{3/2}$，由 **E2** 得知它就是 $2^{3/2+3/2}=2^3=8$。也就是说 $x^2=8$。这并没有完全确定下 x，因为 8 有两个平方根。所以我们通常会采取如下的准则。

E3 如果 $a>0$ 且 b 是实数，那么 a^b 为正数。

再应用上 **E3**，我们就发现 $2^{3/2}$ 是 8 的正平方根。

这并不是对 $2^{3/2}$ 的"真正值"的发现。但这也不是我们对表达式 $2^{3/2}$ 的随意解读——如果我们希望保持规则 **E1**、**E2** 和 **E3**，这就是唯一的可能性。

用类似的办法，我们可以对 a^0 给出解释——至少当 a 不为 0 时。由 **E1** 和 **E2**，我们知道 $a=a^1=a^{1+0}=a^1\times a^0=a\times a^0$。消去律 **M5** 指出，无论 a 取何值，都有 $a^0=1$。对于负指数，如果我们已经知道了 a^b 的值，那么 $1=a^0=a^{b+(-b)}=a^b\times a^{-b}$，由此推出 $a^{-b}=1/a^b$。例如，$2^{-3/2}$ 就等于 $1/\sqrt{8}$。

对数是另一个抽象地来看会变得更加容易的概念。关于对数，我在本书中要说的不多。但如果它确实困扰你，那么你可以消除顾虑，只要了解它们遵循如下三条规则就足以使你去应用对数了。(如果你希望对数是以 e 为底而不是以 10 为底的，只需要在 **L1** 中把 10 替换为 e 即可。)

L1 $\log(10)=1$。

L2 $\log(xy)=\log(x)+\log(y)$。

L3 若 $x<y$，则 $\log(x)<\log(y)$。

例如，要得到 $\log(30)$ 小于 $3/2$，可以应用 **L1** 和 **L2** 得出 $\log(1000)=\log(10)+\log(100)=\log(10)+\log(10)+\log(10)=3$。而由 **L2** 得出 $2\log(30)=\log(30)+\log(30)=\log(900)$，又由 **L3** 得到 $\log(900)<\log(1000)$。因此 $2\log(30)<3$，即得 $\log(30)<3/2$。

在这本书的后面，我还将讨论许多类似性质的概念。试图具体地理解它们会让你感到困惑，但当你放轻松些，不再担心它们是什么并且应用抽象的方法，那这些概念的神秘性就消失了。

第三章
证明

图 9 中画出了五个圆,第一个圆上标出了一个点,第二个圆上标出了两个点,依此类推。连结这些圆上点的所有可能直线段都画了出来,这些线段将圆分割成若干区域。去数一数每个圆的区域数的话,会得到序列 1,2,4,8,16。我们可以立刻辨别出这个序列:似乎圆上每添加一个新的点,区域个数就会加倍,因而 n 个点就分出了 2^{n-1} 块区域——至少在没有三线共点的情况下。

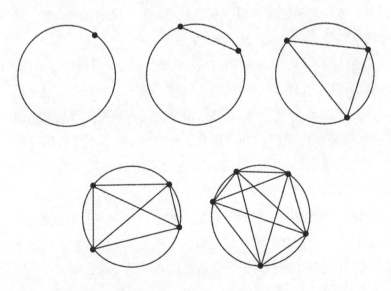

图 9 将圆分割成若干区域

然而，数学家很少会对"似乎"这样的用语感到满意。他们所需要的是**证明**，也就是能够扫清一条论断中所有疑点的论证。可是，这究竟是什么意思呢？尽管我们常常可以立论，在虑及所有**情理之中**的怀疑时认为论断正确；但如果要下结论说我们的论证扫清了一切的疑点，那我们必定要更上一层楼。历史学家能够给出许多例子来说明，一些论断一度被认为毋庸置疑，而后来却被证明是错误的，其中有一部分就是数学方面的论断。为什么当今数学中的定理与此会有所不同呢？我在下面将回答这一问题，给出几个证明的例子来，并从中概括出一些一般性的结论来。

根号 2 的无理性

我在上一章中说到，一个数如果可以写成分数 p/q（其中 p 和 q 是整数）的形式则称为有理数，若不可以则称为无理数。数学中一条著名的证明表明了 $\sqrt{2}$ 是无理数。这项证明阐释了**反证法**这种技术，即通过推出矛盾来证明。

这样的证明以假设要证的结论为**假**开始。这看似有点奇怪，但其实我们在日常对话中经常用到这一技巧。如果你去警局报案，声称目睹了一辆车被人故意破坏，别人指控你自己就是破坏者，那你很可能说："如果是**我**干的，那我根本不可能以这种方式让自己成为目标。"你暂且采纳了你就是破坏者的（非真）假设，以此来说明这是多么荒唐。

我们要去证明 $\sqrt{2}$ 是无理数，那就先假设它是**有理数**，再来表明这一假设会引出荒诞的结论。我在下面将按步骤写出证明，给出比较多的细节，很多读者也许并不需要这么多。

1. 如果 $\sqrt{2}$ 是有理数,那么我们可以找到整数 p 和 q 使得 $\sqrt{2}=p/q$(由"有理数"的定义)。

2. 任意分数 p/q 都能够写成某个分数 r/s,r 和 s 不全是偶数。(分子分母连续除以 2,直到至少有其中一个变成奇数。例如分数 1412/1000 等于 706/500 等于 353/250。)

3. 因此,如果 $\sqrt{2}$ 是有理数,我们就可以找到两个不全为偶数的整数 r 和 s 使得 $\sqrt{2}=r/s$。

4. 如果 $\sqrt{2}=r/s$,则 $2=r^2/s^2$(等式两端平方)。

5. 如果 $2=r^2/s^2$,则 $2s^2=r^2$(等式两端乘以 s^2)。

6. 如果 $2s^2=r^2$,则 r^2 是偶数,即 r 必须是偶数。

7. 如果 r 是偶数,那么存在某个整数 t 使得 $r=2t$(由"偶数"的定义)。

8. 如果 $2s^2=r^2$ 且 $r=2t$,则 $2s^2=(2t)^2=4t^2$,于是得到 $s^2=2t^2$(两端除以 2)。

9. 如果 $s^2=2t^2$,那么 s^2 是偶数,意味着 s 是偶数。

10. 按照 $\sqrt{2}$ 是有理数的假设,我们已经表明 $\sqrt{2}=r/s$,r 和 s 不全是偶数(第 3 步)。我们之后又得到 r 是偶数(第 6 步),s 是偶数(第 9 步)。这是一个明显的矛盾。因为 $\sqrt{2}$ 是有理数的假设会推出明显错误的结论,所以这个假设本身必定是错误的。因此,$\sqrt{2}$ 是无理数。

以上我尽可能使每一步推导都做到明显有理有据,从而使结论无可反驳。但是,我真的完全**没有**给怀疑留下余地吗?若有人愿意跟你打赌,如果找不到两个整数 p 和 q 使得 $p^2=2q^2$ 就给你一万英镑,但如果找到了就处死你,那你愿意接受挑战吗?如果你愿意,又是否会有一点点的不安呢?

第 6 步中包含了 r^2 是偶数则 r 必定是偶数的论断。这看起来相当明显(奇数乘以奇数是奇数),但如果我们要想使 $\sqrt{2}$ 是无理数这一命题**绝对确定无疑**,或许可以再加强一些。让我们把它分解成五个子步骤:

6a. r 是整数,r^2 是偶数。我们要表明 r 必定也是偶数。让我们假设 r 是奇数,然后寻求矛盾。

6b. 因为 r 是奇数,所以存在整数 t 使得 $r=2t+1$。

6c. 于是推出 $r^2=(2t+1)^2=4t^2+4t+1$。

6d. 但是 $4t^2+4t+1=2(2t^2+2t)+1$,这是奇数,与 r^2 是偶数的事实矛盾。

6e. 因此 r 是偶数。

现在步骤 6 完全滴水不漏了吗?可能还没有,因为子步骤 6b 仍需要证明。毕竟,奇数的定义仅仅是非 2 的倍数的整数。为什么每个整数要么就是 2 的倍数,要么就比 2 的倍数多 1 呢?我们可以用以下论据来表明它。

6b1. 如果某个整数 r 是 2 的倍数,或者比 2 的倍数多 1,我们就称它为一个好数。如果 r 是好数,则 $r=2s$ 或 $r=2s+1$,其中 s 也是整数。如果 $r=2s$,则 $r+1=2s+1$;如果 $r=2s+1$,则 $r+1=2s+2=2(s+1)$。不管是两种情况中的哪一种,都有 $r+1$ 也是好数。

6b2. 1 是好数,因为 $0=0\times 2$ 是 2 的倍数,且 $1=0+1$。

6b3. 重复利用 6b1 这一步，我们可以推出 2 是好数，然后 3 是好数，然后 4 是好数，等等等等。

6b4. 因此，所有正整数都是好数，这正是我们要证明的结论。

我们现在完成了吗？大概这一回最不牢靠的步骤要数 6b4 了，因为它是从前一步"等等等等"这样十分模糊的字眼中得到的。步骤 6b3 告诉我们怎样去表明任意**给定的**正整数 n 是好数。但麻烦在于，若按照上面的论证，我们需要从 1 一直数到 n，如果 n 很大的话，这就要花很长时间。如果我们想要说明**所有**正整数都是好数，情况就更糟糕了，看起来这样的论证似乎永无结束之日。

但另一方面，鉴于步骤 6b1 到 6b3 实实在在、确切无疑地给我们提供了一种方法，能够说明任意的 n 都是好数（只要我们有时间），这一反对意见看起来就不合理。实际上它是如此不合理，以至于数学家们采纳了下述原则作为一条公理。

假设关于任意正整数 n 有一陈述 $s(n)$。（在我们的例子中，$s(n)$ 即表示陈述 "n 是好数"。）如果 $s(1)$ 为真，而且 $s(n)$ 为真总蕴含 $s(n+1)$ 为真，那么 $s(n)$ 对任意 n 都为真。

这就是数学归纳法原理，熟悉它的人也简称它为归纳法。通俗地讲，它所说的其实就是，如果你有一列无穷多的陈述序列想要证明，那有一种办法：证明第一条为真，并且每一条都蕴含下一条。

上述几段内容说明了，数学论证中的每一步都可以分解成更小的，因而也更加清晰有据的子步骤。这些小步骤又可以进一步分解为子子步骤，等等。数学中有个根本性的重要事实，那就是这样的过程**最终必然会终止**。原则上，如果不断地将步骤分解为更小的步骤，你最终会得到一条非常长的论证，它以普遍接受的公理开始，仅通过最基本的逻辑原则（例如"若 A 为真且 A 蕴含 B，则 B 为真"）一步步推进，最终得到想要求证的结论。

上一段中我所说的远非显然：事实上，这正是 20 世纪早期的重要发现，很大程度上归功于弗雷格、罗素和怀特海（参见"延伸阅读"）。这一发现对数学产生了深远的影响，因为它意味着，**任何关于数学证明有效性的争论总是能够解决的**。而在 19 世纪，与此形成对比的是，的确存在着关于数学实体问题的真正分歧。譬如，现代集合论之父格奥尔格·康托尔，基于某个无穷集可能比另一个无穷集"更大"这样的思想提出了一些论点。今天人们已经接受了这些论点，但当时的人们却对此产生强烈的怀疑。如今，如果人们对于某个证明的正确性存在分歧，那要么是因为这个证明写得不够详细，要么是因为人们还没有付出足够努力来仔细地理解、检查这个证明。

不过这并不意味着分歧永远不会出现。比如一种常见的情况：某人炮制了一篇极长的证明，在某些地方不够清楚，同时还包含许多小错误，但并非明显可见的根本性错误。要想下结论说这样的论点是否能做到滴水不漏，通常要耗费巨大的工作量，而这样的工作却没有太高的回报。即使是提出证明的人自己可能也不愿意做，生怕会发现他的证明是错的。

虽然如此，争论**在原则上**必然能够解决这一事实的确使数学独一无二。没有任何一个学科像数学一样具有这一特性：有些天文学家仍然固守着宇宙的稳态理论；关于自然选择究竟有多大的解释力，生物学家各自都抱有极为不同的坚定信念；关于意识与物质世界的关系，哲学家们具有根本性的分歧；经济学家也追随着观点截然相反的不同学派，如货币主义和新凯恩斯主义。

理解前面"在原则上"这个短语是很重要的。没有哪个数学家愿意费心写出证明的完整细节——从基本公理开始，仅通过最明显、最易于检查的步骤来推导出结果。即使可行，也并不必要：数学论文是写给经过严格训练的读者的，无须事无巨细详细说明。而如果某人提出一个重要的论断，其他数学家发觉难以理解其证明，他们就会要求作者详细解释。这时，把证明步骤分解为较易理解的小步骤的过程就会开始。同样因为听众都是经过严格训练的，这个过程通常不需要进行太久，只要给出了必要的解释或者发现了其中的错误就可以了。因此，对某个结果的一个证明如果**的确是**正确的，那几乎总会被数学家当作是正确的。

一部分读者可能会萌生这样的问题，我还未触及：为什么我们应该接受数学家提出的公理呢？比方说，如果有人反对数学归纳法原理，我们应当怎样回应呢？大多数数学家会给出如下的答复。首先，所有理解了归纳法的人应该都认为它是显然合理的。其次，公理系统的主要问题并不是公理的**真实性**，而是公理的自洽性和有用性。数学证明实际上所做的正是要表明，由特定前提——如数学归纳法，能够得到特定的结论——如 $\sqrt{2}$ 是无理数。这些前提假设是否正确则是与此完全无关的问题，我们可以安然地把它们留给哲学家。

黄金分割比的无理性

人们在学习高等数学时,走到一个证明的结尾处,通常会经历这样的思考:"我理解每一行是怎样由前一行得到的,但是我却不明白**为什么**这个定理是正确的,人们是怎样想到这个论证的。"我们经常想从证明中得到更多的东西,而不仅仅是确信它的正确性。读过一个好的证明之后,我们会感到它对定理进行了一番阐明,使我们理解了之前所不理解的一些东西。

由于人类大脑有很大一部分是用于处理视觉数据的,我们也就不难理解有很多论证都用到了我们的视觉能力。为了说明这一点,我将给出另一个关于无理性的证明,这一回是所谓的黄金分割比。几百年来,这个数一直使非数学家为之着迷(数学家也着迷,只是程度较轻)。它是具有如下特征的矩形的长宽比:

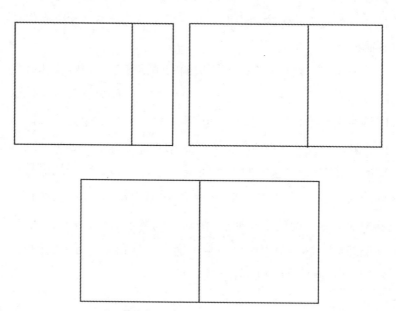

图 10　黄金分割的存在性

从矩形中切掉一个正方形,剩下一个小矩形,它旋转之后恰与原始矩形的形状完全一样。图 10 中的第二个矩形正是这种情况。

为什么这样的边长比是存在的?(数学家已经被训练得惯于提出这类问题。)要回答它可以换一个角度,想象一个小矩形从正方形的一侧生长出来,使整个图形变成一个大矩形。刚开始的时候,小矩形很长很瘦,而大矩形仍然差不多是个正方形。如果我们让小矩形继续生长,直到它自己变成一个正方形,那大矩形的长就是宽的两倍。所以,小矩形最开始的时候比大矩形瘦得多,现在却比大矩形胖(相对于各自的尺寸来讲)。在这两者之间必存在一点,使得两个矩形形状相同。图 10 说明了这样的过程。

还可以用另一种方式来考虑黄金分割比的存在,那就是把它算出来。如果我们把它叫作 x,并假设正方形的边长为 1,那么大矩形的边长就是 1 和 x,小矩形的边长是 $x-1$ 和 1。如果它们的形状相同,那么 $x=\dfrac{x}{1}=\dfrac{1}{x-1}$。两端同时乘以 $x-1$,我们就得到 $x(x-1)=1$,所以 $x^2-x-1=0$。求解这个二次方程,并且知道 x 不是负数,我们就得出 $x=\dfrac{1+\sqrt{5}}{2}$。(如果你在数学上很有造诣,或者深入理解了上一章,你可能会问为什么我如此确信 $\sqrt{5}$ 存在。实际上,这第二种论证正是要把几何问题归结于一个等价的代数问题。)

得出了比例 x 存在之后,让我们来考虑对边长为 x 和 1 的矩形进行如下的操作。首先,从中切掉一个正方形,由黄金分割比的定义,剩下的小矩形与原来的矩形形状相同。然后再不断地重复这一基本操作,得到一系列越来越小的矩形,每一个的形状都与之前的形状相同，因而长宽比都是黄金分割比。很显然，这一过程永远不会终止。（参见图 11 的第一个矩形。）

现在让我们对长宽比为 p/q 的矩形施行同样的操作，其中 p 和 q 都是整数。也就是说，这个矩形与边长为 p 和 q 的矩形形状相同，因而可以被分成 $p \times q$ 个小正方形，如图 11 中的第二个矩形所示。如果我们从这个矩形的一端切掉一些大正方形会怎么样呢？如果 q 小于 p，那么我们切掉一个 $q \times q$ 的正方形，得到一个 $q \times (p-q)$ 的矩形。我们可以继续切掉下一个正方形，依此类推。这个过程会永远持续下去吗？不会。每次切掉的正方形都是小方格的整数倍，我们不可能操作多于 $p \times q$ 的次数——因为一开始就只有 $p \times q$ 个小方格。

我们表明了如下两个事实。

1. 如果矩形的边长比是黄金分割比，那么我们可以一直从中切掉正方形，没有尽头。
2. 如果矩形的边长比是某对整数 p 和 q 的比值 p/q，那么就不能永不停止地从中切掉正方形。

于是我们可以得到，比值 p/q 不是黄金分割比，无论 p 和 q 取什么值。换言之，黄金分割比是无理数。

仔细思考上面这个证明，你最终会发觉，虽然乍看它和之前对 $\sqrt{2}$ 是无理数的证明很不一样，但实际上也并无太大区别。不过，我们将它呈现出来的方式确实不同——而且对很多人来说，这种方式更有吸引力。

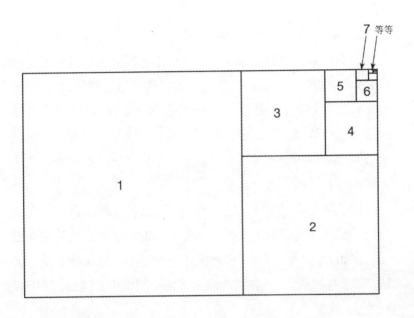

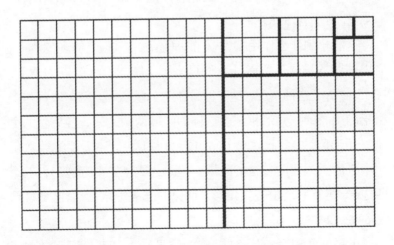

图11 从矩形中切掉正方形

圆的分割

既然我已经谈到了数学证明的本质，现在让我们回到本章开始时的问题。圆上有 n 个点，我们用直线将这些点两两连结起来，希望能够表明这些直线所分割出的区域总数是 2^{n-1}。对于 n 为 1，2，3，4 或 5 的情形，我们已经看出这是对的。为了一般性地证明这个陈述，我们很想找到一种令人信服的**推理**，能够说明边界上每增加一点区域总数就增加一倍。这样的推理会是什么样的呢？

脑子里无法立刻蹦出些什么来，那么我们可以通过观察被分割的圆的图形来着手，看看能否从中发现某种模式并提炼概括。例如，圆上有三个点，就产生了三个外围区域和一个中心区域。圆上有四个点，就有四个外围区域和四个中心区域。圆上有五个点，就有一个中心五边形，五个三角形从中心指向外围，五个三角形嵌入这个五角星之中，于是又形成一个五边形，最后还有五个外围区域。因此，我们可以把 4 看作 3+1，把 8 看作 4+4，把 16 看作 5+5+5+1。

这能起到什么作用吗？我们似乎还没能掌握足够多的例子，看出清楚的模式，所以让我们再来画出圆上有六个点的情形。图 12 显示了这种情况。这一回外围有六块区域。每一块都与一个指向中心的三角形相邻。这样的区域两两之间，各夹着两个小三角形区域，至此共有 6+6+12=24 块区域，还需要继续数位于中心的六边形内包含的区域。里面分成了三个五边形，三个四边形，还有一个中心三角形。所以看起来很自然地，区域总数是 6+6+12+3+3+1。

可是好像出了点什么错，因为结果是 31。我们哪里疏忽了

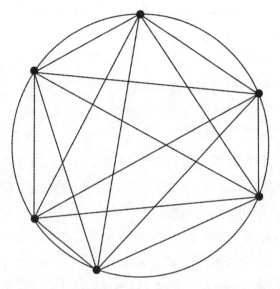

图 12　圆的分割

吗？事实上并没有：正确的序列开头几项是 1, 2, 4, 8, 16, 31, 57, 99, 163。其实只要稍加深入思考，我们就能看出，区域总数**不可能**每次都翻倍。在刚开始时就有点麻烦，圆上 0 个点得到的区域总数是 1 而不是 1/2，所以加上第一个点的时候区域总数就不是翻倍。尽管这种例外情形时常发生在 0 上，大多数数学家还是会认为这是个麻烦。但是，当 n 是个比较大的数时问题会更加严重，这时 2^{n-1} 显然会特别大。比如当 $n=20$ 时 2^{n-1} 是 524 288，当 $n=30$ 时，结果是 536 870 912。在圆的边上画 30 个点就会把圆分割成超过 5 亿个不同区域，这可能么？当然不可能。想象一下，在一块地上画一个大圆，在圆上打 30 个间隔不一的桩子，用很细的线把桩子两两相连。结果得到的区域数量确实比较大，但也不会大到难以想象的程度。如果圆的直径是 10 米，把它分成 5 亿块，平均每平方厘米里就会有超过600块区域。这个圆一定密密麻麻布满了线,但周围只有30个点,实际情况显然不是这样的。

我前面说过,数学家对待"显然"这样的词非常慎重。但在这个例子中,我们的直觉能够以坚实的论证来支持,归结如下。如果圆被分割成数量巨大的多边形区域,这些区域之间必然有大量的顶角。每个顶角处都是两条线的交点,和这个交点相联系的桩子,也就是两条线在圆上的端点有4个。我们在圆上选取这样的4个桩子,第一个桩子有30种可能的选择,第二个有29种,第三个有28种,第四个有27种。这意味着,选取4个桩子的方式共有 $30 \times 29 \times 28 \times 27 = 657\ 720$ 种,但是这样就没有考虑到,如果以不同的顺序选取了同一组桩子,就重复列入了相同的交点。选出同一组4个桩子共有 $4 \times 3 \times 2 \times 1 = 24$ 种不同方式,考虑到这一点,我们就能够算出交点总数是 $657\ 720/24 = 27\ 405$ 个,根本不可能像536 870 912块区域的顶角数量那样巨大。(实际上,30个点分割出的区域的真正数量是27 841。)

在这个让我们引以为戒的故事中,包含了很多与证明数学陈述相关的重要教训。最明显的一个就是,如果不去小心地证明你所说的话,那你就有说错的危险。一个更积极一点的寓意是,如果确实努力去证明一个陈述,那你将能以全然不同而且更有意思的方式理解它。

毕达哥拉斯定理

毕达哥拉斯的著名定理所讲的是,假设一直角三角形的三边长为 a、b 和 c,其中 c 是斜边长(直角所对的边),则 $a^2+b^2=c^2$。这个定理有若干种证明,其中有一种特别简短,而且很容易理解。它差不多只需要下面两幅图即可。

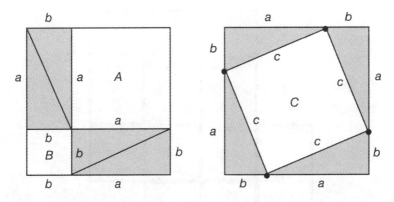

图 13　毕达哥拉斯定理的简短证明

图 13 中，我标为 A、B、C 的正方形，边长分别为 a、b、c，因而面积分别为 a^2、b^2、c^2。因为四个三角形的移动并没有引发面积的改变，也没有使它们重合，所以两幅图中，大正方形去掉四个小三角形所得的面积应当是相同的。但在左图中，这个面积是 a^2+b^2，而右图中则是 c^2。

缺角正方形网格的铺地砖问题

这里有个著名的难题。画八横八纵正方形网格，去掉相对的两个角。你能用多米诺骨牌形状的地砖——每一块正好覆盖两个相邻方格，把剩余部分覆盖吗？我在图中(图 14)表明，如果用四横四纵来代替八横八纵，你是办不到的。假设你决定用一块地砖覆盖我图中标为 A 的区域，那么容易看出，你必须还要把地砖放到 B、C、D 和 E 的位置上，剩下一个小方格无法覆盖。既然右上角的格子无论如何总要覆盖住，而仅剩的另一种覆盖的方式也会导致类似的问题(通过位置的对称关系)，所以覆盖整个图形是不可能的。

如果我们用五来代替四，网格的铺设仍然是不可能的，原因

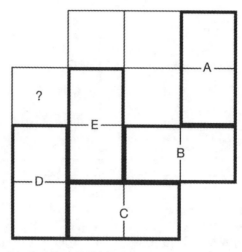

图 14　缺角正方形网格的铺地砖问题

很简单,每块地砖占两个方格,而小方格总数是 23——是个奇数。但是 $8^2-2=62$ 是个偶数,所以我们不能把这样的论证用于八横八纵的方格。另一方面,若想找到一种类似于刚才四横四纵情况中我给的证明,你很快就会放弃,因为你所需要考虑的可能情况实在太多了。那对这个问题应该怎样入手呢?如果你从没接触过这个问题,我强烈建议你在继续阅读之前先尝试求解一下,或者暂时跳过下一段,因为如果你解决了它,你将能够对数学中的愉悦感有很好的理解。

对于无视我建议的读者——经验表明这些人占大多数,有一个词几乎可以代表全部证明:国际象棋。国际象棋的棋盘是八横八纵的方格。每个小格交错地填上黑色和白色(就象棋游戏本身来考虑似乎并不是必要的,不过能够使视觉看起来更轻松)。两个相对的顶角方格颜色是相同的。如果它们都是黑色的,那么一旦把它们去掉,剩下的棋盘就有32个白格子和30个黑格子。每一块骨牌只能覆盖两种颜色的方格各一块,所以一旦你放进了30块骨牌,无论是怎么放的,最终都必然剩下两个白方格,它们是无法覆盖的。

这个简短的论证极好地表明,证明何以能够不仅仅保证陈述的正确性。例如,四乘四方格去掉相对两角无法被覆盖,这条陈述我们现在有了两种证明。一种是我前面给出的,另一种是四乘四版本的象棋盘论证法。这两种证明都得到了我们想要的结果,但只有第二种给了我们一种关于无法覆盖的类似**推理**的东西。这样的推理能够立即告诉我们,一万乘一万的方格去掉相对的两角也是无法覆盖的。第一种论证则只能告诉我们四乘四的情况。

第二种论证有个值得瞩目的特征,它完全依赖于一种思想,这种思想虽出人意料,但一经理解便显得非常自然。人们经常很困惑,为什么数学家有时会用"优美"、"漂亮"甚至"绝妙"来形容一些证明。这样的例子就让我们对其含义有了一点理解。音乐也能够提供一个有用的类比:一段乐曲刚开始可能沿意想不到的和声方向行进,过后却感觉到非常完美恰当,或者一段管弦织体呈现出整体大于部分之和的境界,其方式我们还无法全然理解——每当这些时候我们就会为之陶醉。在数学证明中,有突如其来的启发,有出人意料却自然而然的思想,还有引人入胜、有待进一步发掘的暗示,这些都能够给我们带来类似的愉悦感。当然,数学的美不同于音乐的美,可音乐的美同样也不同于绘画的美、诗歌的美、姣好面容的美。

三条看似显然实则需要证明的陈述

较高等的数学中,有一点让很多人感到费解:其中有一些定理看上去非常显然,简直无须证明。遇到这样的定理时,人们常常会问:"如果**这**都不算显然,那还有什么才算呢?"我一位先前的同事对此给了一个很好的回答:如果脑子里立刻就有证明,那么这条陈述才是显然的。在本章的剩余部分,我将给出三条陈述作为例子,它们看上去都是显然的,却无法通过这样的检验。

  1. 算术基本定理的内容是,每个自然数有且仅有一种被写为素数乘积的方式,不考虑先后顺序。例如,$36=2×2×3×3$,$74=2×37$,再如101本身就是一个素数(在这样的语境下,它就是单个素数的"乘积")。观察过这些较小自然数,人们马上就会确信,根本不可能有两种**不同**方式把一个数表示为素数乘积。这就是这个定理看似几乎无须证明的主要论点。

  但它真的有那么明显吗?$7,13,19,37,47$都是素数,那么,如果算术基本定理是显然的,$7×13×19$不等于$37×47$也应该是显然的。我们当然可以检验出,这两组数确实不同(任何数学家都会告诉你,其中一个比另一个更有意思①),但这并不说明它们**显然**会不相等,也不能解释为什么我们不能另外找到两种素数乘积得到相同的结果。实际上,这个定理并没有很简单的证明方法。如果你脑子里立刻就有了证明,那你的脑子一定很不寻常。

  2. 设想你用一段正常的绳子打了一个宽松的结,然后把绳子两端熔合在一起,得到了如图15所示的形状即数学家所称的三叶结。在不把绳子切断的情况下,有没有可能把这个结解开呢?当然不可能。

  但为什么我们倾向于说"当然"呢?我们有没有立刻就想到一种论证?可能有——看起来,任何解开结的努力似乎都难以避

---

① 英国数学家哈代回忆印度数学家拉马努金时谈到,某次他去医院看望拉马努金,提到路上搭乘的出租车的车牌号为1729,并无特殊之处。拉马努金说:"你错了。实际上,这是能以两种方式表示为两数立方和的最小自然数。"

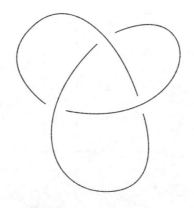

图15　三叶结

免地会使得这个结更为繁复。然而，要把这样的直觉转换为有效的证明却是很难的。所谓显然，只不过是没有**简单**的方法解开这个结。难点在于需要排除所有的可能性，说明**先把结变复杂**再最终解开它也是不可能的。应当承认，这看起来不太可能，但是，数学中的确存在着这样的现象，甚至在日常生活中都是存在的：比如，为了把房间收拾干净，常常在开始时有必要先把它弄得更乱一些，而不是把所有东西都塞进橱里。

3. 平面中的一条**曲线**，是指笔尖不离开纸面的情况下，你所能画出的任何东西。如果曲线永远不和自己相交，就称它为**简单的**，如果它最终回到起点处，就称它为**闭合的**。图16显示了这些定义在实际图形中的反映。图中第一条曲线既是简单的也是闭合的，在平面上围出一块区域，即曲线的**内部**。很明显，所有简单闭合曲线都将平面分成两部分：曲线里面和曲线外面（如果把曲线本身也看作一部分，就是三个部分）。

这真的有那么明显吗？当曲线不太复杂的时候的确如此。但如图17所示的曲线呢？如果你在靠近图形中心处选一点，点在曲线里面还是外面并不显然。你可能想，大概的确不太显然，可

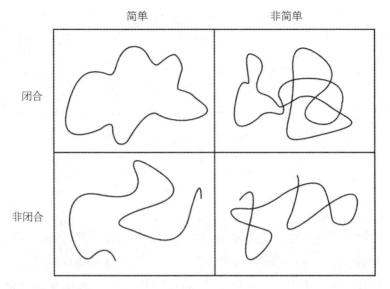

图16 四种曲线

图17 黑点在曲线里面还是外面？

它终归必定会**有**一个里面和一个外面，哪怕曲线很复杂，在视觉上很难辨别。

应该怎样来确证这样的看法呢？我们可能会尝试用如下的办法来区别里面和外面。我们暂时假设，确实存在曲线的里面和外面，那么每次穿过曲线的时候，必定要么是从里面穿到外面，要么是从外面穿到里面。于是，如果你想确定点P是在里面还是在外面，你只需从P出发画一条线，一直画到远离曲线的另一点Q，也就是使Q明显位于曲线外面。如果画的这条线和原曲线相交奇数次，那么P位于曲线里面，否则位于曲线外面。

这种论述的问题在于，它把许多事情当作理所当然的了。例如，如果你从P出发画**另一条**线，画到曲线外另一点R，你怎么知道结果不会不同呢？（结果不会不同，但这需要证明。）实际上，任意简单闭合曲线都有里面和外面这一命题，正是一个著名的数学定理，称作若尔当曲线定理。无论它看起来多么显然，它都需要证明，而且任何我们已知的证明都非常困难，远远超出了本书的论述范围。

第四章
极限与无穷

在上一章中，我努力指出：原则上，数学证明如何可以完全形式化。如果从特定的公理开始，遵循特定的规则，最后以有趣的数学陈述结束，那么这样的陈述就可以当作定理接受，否则就不能被视为定理。这种思想，即从少数几条公理出发演绎推导出许多复杂的定理，可以追溯到欧几里得。欧几里得只用了五条公理就建立起几何学的主要体系。（关于他的公理，我们将在第六章中讨论。）有人可能提出这样的问题：为什么直到20世纪，人们才认识到这样的思想可以应用于整个数学系统当中呢？

主要的原因可以被归结为一个词：无穷。出于种种原因，无穷这一概念在数学中必不可少，但却很难严格化。在本章中，我将讨论三条陈述。其中每一条乍看起来都普普通通，但经过仔细的考察，会发现最终都涉及到无穷。随之就产生了困难，本章的主要内容就是如何处理这样的困难。

1. 2的平方根约为1.414 213 56

上面这条简单陈述仅仅在说一个不大的数差不多等于另一个数，哪里涉及到无穷了呢？答案就藏在"2的平方根"这个短语里面，这个短语隐含假设了2的确**存在**一个平方根。要想透彻地理解这条陈述,我们就必须问一问,2的平方根究竟是个什么样的对象。于是无穷就来了:2的平方根是一个无穷小数。

注意,下面这条紧密相关的陈述就不涉及无穷的问题:1.414 213 56的平方约为2。这条陈述完全是有限的,但是看起来所谈论的基本上是同样的事情。我们稍后就能看到,这一点至关重要。

说一个无穷小数的平方等于2,这是什么意思呢?在学校里,我们学过有限小数如何相乘,但从没学过无穷小数的乘法——不知出于什么缘故,我们仅仅假设它们能够参与加法和乘法,而不去深究。这运算应该如何完成呢?为了看看会出现何种困难,我们先来考虑加法。当我们把两个有限小数相加时,比方说2.3859加3.1405,我们将一个数写在另一个数下面,从右向左将对应的数位相加。我们从最末端的数位开始做加法——9加5,得到14,于是写下4,并进1。接下来,我们再加倒数第二位——5加0,以及进上来的1,得到6。依此类推,我们就得到了最终的结果5.5264。

现在假设有两个无穷小数。我们不可能从最右端开始,因为无穷小数根本没有最后一位。那么我们要如何把它们相加呢?有一个明显的回答:从左端开始。然而这么做是有缺陷的。譬如,我们再来加有限小数2.3859和3.1405,先加2和3,得到5。接下来再加3和1,得到4,很不幸,这是不对的。

这个错误给我们造成了一点麻烦,不过只要我们保持勇气继续下去倒也没什么大不了的。接下来要相加的两个数字是8和4,我们可以在这第三位上记下一个2,然后将上一位的4改为5加以**修正**。这个过程还会出现,在第四位我们会先写下一个5,然后修正为6。

注意，这样的修正有可能会在写下该位结果的很久之后才出现。例如，我们要计算 1.355 555 555 555 555 557 3 加 2.544 444 444 444 444 445 2，会首先写下 3.899 999 999 999 999 99，但再往下算一位，即7加5时这一连串的9就不得不修正。就像一连串多米诺骨牌一样，这些9一位接一位地变为0。不过这样的计算方法仍然是行得通的，最后得到计算结果 3.900 000 000 000 000 002 5，它让我们能够对如何将无穷小数相加赋予意义。不难看出，没有哪一位数需要修正一次以上，所以如果我们有两个无穷小数，那么两数之和的第53位（打个比方），要么是我们在上述过程的第53步写下的那个数，要么是之后修正的那个数——如果有必要修正的话。

我们想要对这一论断赋予意义：存在一个无穷小数，其平方为2。要做到这一点，我们必须首先看一看这个无穷小数是怎样产生的，然后再去理解让它和自己相乘意味着什么。可以预料到，无穷小数的乘法会比加法更为复杂。

首先，有一个产生这个小数的很自然的办法。这个数肯定在1和2之间，因为 $1^2=1$，小于2，而 $2^2=4$，大于2。如果你一一计算 $1.1^2, 1.2^2, 1.3^2$，一直到 1.9^2，你会发现 $1.4^2=1.96$ 小于2，而 $1.5^2=2.25$ 大于2。所以 $\sqrt{2}$ 必然在1.4和1.5之间，因此它的小数展开一定是以1.4开头。现在假设你已经以此办法算出 $\sqrt{2}$ 的前八位数字是1.414 213 5。你可以继续如下的计算，得到下一位数字是6。

$$1.41421350^2=1.9999998236822500$$
$$1.41421351^2=1.9999998519665201$$
$$1.41421352^2=1.9999998802507904$$

$$1.41421353^2=1.9999999085350609$$
$$1.41421354^2=1.9999999368193316$$
$$1.41421355^2=1.9999999651036025$$
$$1.41421356^2=1.9999999933878736$$
$$1.41421357^2=2.0000000216721449$$

重复这一过程,你想算出多少位都可以。虽然这样的计算永远无法结束,但是你至少有了一种清晰明了的方法,能给定小数点后第 n 位的数字,无论 n 取何值:小数点后有 n 位数,平方不大于2的最大小数,$\sqrt{2}$ 的小数点后第 n 位数字就与它的最后一位相同。例如,在所有小数点后有两位的小数中,1.41是平方不大于2的最大的一个,所以 $\sqrt{2}$ 展开到小数点后两位是1.41。

让我们记最终得出的无穷小数为 x。是什么让我们能够坚信 $x^2=2$ 呢?我们可能会给出如下的理由。

$$1^2=1$$
$$1.4^2=1.96$$
$$1.41^2=1.9881$$
$$1.414^2=1.999396$$
$$1.4142^2=1.99996164$$
$$1.41421^2=1.9999899241$$
$$1.414213^2=1.999998409469$$
$$1.4142135^2=1.99999982368225$$
$$1.41421356^2=1.9999999933878736$$

上述算式列表显示了，$\sqrt{2}$的小数展开位数越多，自身相乘得到的小数点后的数字9就越多。因此，如果把$\sqrt{2}$完整地展开到无穷多位，我们应该得到无穷多个9，而1.99999999…（9无限循环）等于2。

这样的论述会导致两个困难。其一，为什么1.999999…等于2？其二，也是更严重的一个问题，"完整地展开到无穷多位"是什么意思？这是我们首先想努力搞懂的问题。

为了解决第一个困难，我们必须再次搁置一切直觉。在数学中，人们的确普遍认为1.999999…等于2，但这个事实并不是经由某种形而上学的推理过程发现的。相反，它只是种**传统习惯**。但这也绝不是毫无理由的传统，因为如果不采纳它，我们就必须发明怪异的新对象，或者抛弃一些熟悉的算术规则。比如，如果你坚持1.999999…不等于2，那么2-1.999999…是什么呢？如果这是0，那么你也就抛弃了一条有用的规则：若x-y=0，则x必定与y相等。如果这不是0，那这个数就没有通常的小数展开（否则，你从2中减掉这个数，得到的就不是1.999999…，而是较小的别的数），所以你必须创造一个新的对象，诸如"0加小数点后无穷多个0，**之后**是个1"。如果做这样的事，那你的麻烦才刚刚开始。这个神秘的数自乘是什么东西呢？小数点后无穷多个0，之后再来无穷多个0，**之后**是个1？如果你用它乘以10呢？你得到的是不是"无穷多减1"个0，之后是个1？1/3的小数展开是什么呢？用它乘以3，答案是1还是0.999999…？如果你遵循了惯用的传统，这一类难以对付的问题就不会产生了。（虽难以对付，但也并非全然不可能：在1960年代，亚伯拉罕·罗宾逊发现了对"无穷小"数的一个条理清晰的定义，但正如其名"非标准分析"所暗指的,这套理论还没有成为数学界的主流部分。)

第二个困难是更实实在在的困难,但它也是可以绕过去的。我们不去试图想象对无穷小数进行长乘计算,而仅仅将陈述 $x^2=2$ 解释为,x 展开位数越多,平方的结果就越接近2,一如我们已经观察到的。为了更确切地说明这一点,假设你仍然坚持想要一个数,它的平方以1.9999…开始。我会向你建议1.414 21这个数,它是由 x 的前几位给出的。因为1.414 21很接近1.414 22,我预期它们的平方也会很接近(这很容易证明)。但根据我们选择 x 的办法,必然有 $1.414\ 21^2$ 小于2而 $1.414\ 22^2$ 大于2。因而这两个数都很接近于2。只要检查一下:$1.414\ 21^2=1.999\ 989\ 924\ 1$,于是我们就找到了满足条件的一个数。如果你想要一个新的数,其平方开始几位是

1.99999999999999999999999999999999999999…

那我可以用同样的办法,只不过多取 x 几位而已。(事实上,如果你想要 n 个9,那取到小数点后 $n+1$ 位就足够了。)无论你想要多少个9,我都可以办到,这一事实正是无穷小数 x 自乘等于2的含义。

请注意,我所做的是"驯服"无穷,只是将涉及无穷的陈述单纯解读为一种生动的简化,其所指的乃是一条不涉及无穷的累赘得多的陈述。关于无穷的简洁陈述是"x 是平方等于2的无穷小数"。可以大致翻译成:"有这样一种规则,对任意 n,它能够切实地给出 x 的前 n 位数字。这使我们能够算出任意长的有限小数,它们的平方接近于2,只要算得足够长,想要有多接近就能有多接近。"

我是不是在说，$x^2=2$这个外表简单的陈述的真正意思其实非常复杂呢？某种意义上我的确是这个意思——这条陈述确实隐含了复杂性；但从更重要的意义上讲，我并非是指这个意思。固然，在不提及无穷的情况下，要定义无穷小数的加法和乘法是很难的，而且我们还必须检查这个复杂的定义遵从第二章中列出的那些规则，诸如交换律和结合律。但是，一旦给出了这个定义，我们就可以再次无拘无束地进行抽象思考。关于x，重要的是它的平方等于2。关于"平方"一词，重要的是它的含义以某种乘法定义为基础，这种定义遵循着恰当的规则。x的第一万亿位是什么并不是真正紧要的，乘法的定义有多复杂也不真正紧要。

2. 我们经过那个路灯柱时时速为40英里

假设你坐在一辆正在加速的小轿车里，看着速度表的指针稳定地从时速30英里移动到时速50英里。我们很容易说，在某个时刻——恰在指针扫过40的那一瞬间——汽车以时速40英里的速度行驶。在那个时刻以前，车速小于它，在之后车速大于它。但是，说仅在某个瞬间车速为每小时40英里，这是指什么意思呢？如果汽车没有加速，那我们可以测量它一小时走过多少英里，从而得到它的速度。（更具操作性的是，观察它在30秒内走过多远，再乘以120。）但是，这种办法对加速的汽车显然无效：测量它在一段时间内走过多远，只能算出这段时间内的**平均**速度，并不能得到特定瞬时的速度。

如果我们能够测量在**无穷小**的一段时间内汽车走了多远，那问题就不存在了，因为这样速度就没有时间发生变化。如果这段时间是t小时，t是个无穷小的数，那么我们可以测量汽车在t

小时走过多少英里，记测量结果为s，它当然也是一个无穷小的数，用它除以t就得到了汽车的瞬间速度。

这种荒唐的幻想引起的问题相当类似于前面遇到的情况，当时我们简要地考察了1.999999…可能不等于2的思想。t是0吗？如果是，那很明显，s必然也是0(没有时间流逝，汽车不能走过任何路程)。但是，无法通过0除以0来得到清楚明白的答案。如果t不是0，那在这t小时之内汽车就会加速，从而导致测量无效。

要想理解瞬时速度，我们就要利用这样的事实：如果t非常小——比如百分之一秒，则汽车没有时间去加速**太多**。设想对于某个时刻，我们不去试图计算确切的速度，改而用一个较准的估计来代替。这样，如果我们的测量装置是精确的，我们就能知道汽车在百分之一秒内走了多远，然后把这段距离乘上一小时内百分之一秒的数量，即乘以360 000。这个答案并不会特别准确，但既然汽车在百分之一秒内加速不多，这个答案就给了我们一个相当接近的近似值。

这个情况让我们回想起$1.414\,213\,5^2$相当近似于2这一事实，于是类似地，我们可以不用再操心无穷，或者说是本例中的无穷小。假设我们不是在百分之一秒内测量，而改为在百万分之一秒内测量，那汽车的加速就更少了，因而得到的答案会更加精确。这种观察让我们有了一种办法，可以把陈述"汽车**此刻**的行驶速度是40英里每小时"转换成以下这条更复杂的、不涉及无穷的陈述："如果你给定了允许的误差范围，那么只要t足够小(远小于1)，我就可以测出汽车在t小时内走过的英里数，再除以t，得到的结果将很接近于40英里每小时，误差在你规定的范围之内。"例如，如果t足够小，我可以保证使我的估计值位于39.99和40.01之间。如果你要求我的答案精确到误差小于0.0001，那么

我可以使 t 变得更小。只要它足够小，我总可以满足你要求的精度。

我们再一次把一条涉及无穷的陈述，视为对一条更复杂的、关于近似的命题的方便表达。另一个更具提示性的词语是"极限"。一个无穷小数是一列有限小数的极限，瞬时速度是通过测量越来越短的时间内走过的距离所得近似值的极限。数学家经常谈论"在极限时"或者"在无穷时"的情况如何，但他们都很明白，他们并没有把这种说法完全当真。如果强迫他们说出确切意思，他们就会转而谈论近似。

3. 半径为 r 的圆面积为 πr^2

无穷可以借由有穷的情况来理解——人类对这一点的认识就是19世纪数学的伟大胜利之一，不过这种思想的根源能够追溯到更早。我的下一个例子是讨论圆面积的计算，就将使用阿基米德在公元前3世纪所提出的论证。不过，在开始计算以前，我们应当明确要算的是什么，这个问题并没有乍看上去那么简单。什么**是**面积？当然，它是图形里**某种东西的量**（二维的东西），但怎样才能精确表达呢？

不论它是什么，对特定形状来说，它看起来当然是很容易计算的。例如，如果矩形的边长为 a 和 b，那它的面积就是 ab。任何直角三角形都可以被看作把矩形沿一条对角线切下一半所得，因此它的面积是对应矩形面积的一半。任意三角形都可以分割成两个直角三角形，任意多边形都可以分割成一系列三角形。因此，要算出多边形的面积并不会太难。不必去烦恼我们算的到底是什么，我们只要单纯地将多边形的面积**定义**为此计算结果即可（只要我们能够确信，将多边形切割成三角形的不同方法不会导致不同的答案)。

当我们考虑边缘为曲线的图形时,问题就来了。将圆切割成若干个三角形是不可能的。那么当我们说圆的面积是 πr^2 时,我们谈的是什么呢?

这又是抽象方法大有用处的一个例子。让我们不要关注面积**是**什么,而是关注面积能够**做**什么。我们要对这个说法作一些澄清,因为面积看上去做不了什么事情——它只不过是就在那里而已。我所说的意思是,我们应当关注,关于面积的任何合理概念所应具有的一些属性。下面列出了五条。

Ar1 移动图形,图形面积不变。(更正式的说法:两个全等的图形面积相等。)

Ar2 如果一个图形完全包含于另一个之内,那么第一个的面积不大于第二个。

Ar3 矩形的面积通过它的长宽相乘得到。

Ar4 将图形切成若干部分,则各部分面积之和等于原图形的面积。

Ar5 图形向各方向扩张为原来的2倍,则图形面积变为原来的4倍。

如果回过头来看,你会发现我们利用了属性 **Ar1**、**Ar3** 和 **Ar4** 来计算直角三角形的面积。属性 **Ar2** 看起来非常显然,几乎无须一提,不过作为公理正应如此,而且我们后面还会看到它会多有用。属性 **Ar5** 尽管很重要,其实并不需要作为一条公理,因为它能从另外几条中推导出来。

我们怎样利用这些属性来谈论圆面积的含义呢?截至目前,

这一章所要向读者传达的信息就是，不要考虑一步到位的定义，考虑面积的**近似**可能会有收获。我们可以采取下面这种办法轻易做到。想象一个图形画在一张布满细密方格的纸上。由属性**Ar3**，我们知道这些方格的面积(因为正方形只是特殊的矩形)，所以我们可以数一数有多少正方形完全处于图形以内，以此来估计图形的面积。举例来说，图形中包含了144个小方格，那么图形的面积至少是144乘以单个方格面积。注意，我们所真正计算的面积仅是144个方格所构成的图形的面积，此面积由属性**Ar3**和**Ar4**可以很容易地确定。

对于图18所示的图形，这并不能给出正确的答案，因为还有若干个方格部分位于图形之内、部分位于图形之外，我们还没有计入它们的面积。不过，有一种改进估值的明显方法，就是把

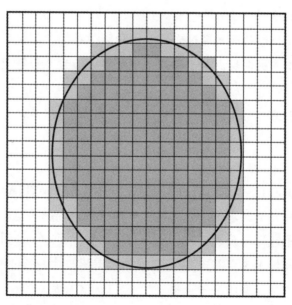

图18　曲线图形面积的逼近

每一个小方格分成四个更小的方格，改用它们来计算。和之前一样，又有一些方格部分位于图形外、部分位于图形内，但这回我们已经多包含进了一些完全在图形之内的小方格。一般来说，方格划分得越密，我们在计算中能够计入的原图面积也就越多。我们发现（这一点并没有看上去那么显然），随着把网格画得越来越密，方格越来越小，计算的结果也就越来越接近于某个数，就像 $\sqrt{2}$ 近似值的平方越来越接近于 2 一样，于是我们就将这个数**定义**为这个图形的面积。

所以，从数学的角度来讲，某个图形的面积为一平方码的含义是这样的：在特定的容许误差范围之内，无论多小，总可以找出充分密集的方格，由位于图形内的小方格来近似计算，得到的计算结果与一平方码之差少于给定误差。（在脑子里某处可能有这样的想法——但已经被压制住了："在极限下"可以使用无穷多无限小的方格来得到确切的答案。）

还有另外一种说明方法，也许能够说得更清楚。如果一个曲线图形的确切面积是 12 平方厘米，要求我用方格法来表明这一点，这个任务是不可能完成的——我需要用无穷多的方格才能做到。但是，如果你给出任何**其他**不同于 12 的数字，比方说 11.9，那我都可以用一套方格来确定地证明它的面积**不是**这个数：我只需使网格足够密，让没有计入的部分小于 0.1 平方厘米即可。换句话说，在不涉及无穷的情况下，我不去证明它的面积是 12，而是满足于证明它的面积不是别的任何数。图形的面积是我所不能证伪的那个数。

这些思想给了面积一个令人满意的定义，但仍给我们遗留下一个问题。我们要怎样来说明，如果采用上述过程来估计半径为 r 的圆的面积，估计值就会越来越接近 πr^2 呢？对于大多数图

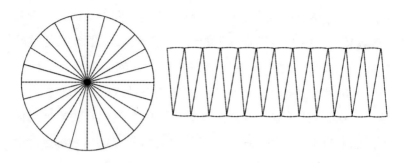

图19 阿基米德说明圆面积为πr^2的方法

形来说,答案是必须使用到积分,我这本书中不会去讨论它。但对于圆来说,正如我之前所提到的,我们可以使用阿基米德的绝妙论证。

图19表示的是,将圆切成一片一片,再组合成接近于矩形的形状。因为每一小片都很窄,所以矩形的高度大约为圆的半径r。同样因为每一片很窄,这个准矩形的上边和下边都近似于直线。上下两边各用了圆的周长的一半,由π的定义知圆的周长为$2\pi r$,则两边长度均近似为πr。因此,准矩形的面积是$r \times \pi r = \pi r^2$——至少是近似如此。

当然了,事实上准矩形的确切面积就是πr^2,因为我们所做的只不过是把圆切开,再移动各部分而已,但是我们目前还不知道这一点。到现在为止,这个论证可能已经使你确信了,但其实还没结束,因为我们还必须表明,随着切开的片数越来越多,上述近似值会越来越接近πr^2。一种非常简明的方法是利用两个正多边形,一个恰好包含在圆内,一个恰好包含圆。图20用六边形说明了这一点。内接多边形的周长小于圆的周长,而外切多边形的周长大于圆的周长。这两个多边形都可以切成多个三角形,再拼合成平行四边形。一点简易的计算就能够表明,较小的

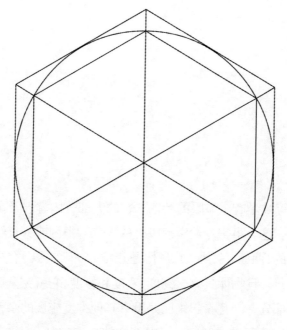

图20　由多边形逼近圆

平行四边形面积小于 r 乘以内接多边形周长的一半，因而小于 πr^2。类似地，较大的多边形面积大于 πr^2。如果切出的三角形数量足够大，那么这两个多边形面积之差就会越来越小，想要多小就有多小。因为圆总是包含着小多边形，并被大多边形所包含，所以它的面积必然恰为 πr^2。

第五章
维度

高等数学有个显眼的特征，它的大部分内容涉及高于三维的几何。这个事实令数学家以外的人感到很困惑：直线和曲线是一维的，曲面是二维的，形体是三维的，但怎么会有东西是四维的呢？一旦物体具有了高度、宽度和厚度，它就完全填充着空间的一部分，而且看起来已经没有其他维度施展的余地了。有人会提出第四维是时间——在特定情境下这是个不错的回答，比如在狭义相对论中。但它并不能帮助我们理解诸如二十六维乃至无穷维的几何，而这些几何在数学上又都很重要。

高维几何又是一例最好从抽象角度来理解的概念。让我们不去担心二十六维空间的**存在**等等，而去考虑它的**性质**。你可能会疑惑：这东西连是否存在都不确定，怎么可能考虑它的性质呢？不过这样的担心很容易解决。如果丢弃"这东西"一词，那么这个问题就变成了：连**拥有**这种性质的对象是否存在都不知道，怎么可能考虑这种性质呢？但这个问题一点都不困难。比如说，我们完全可以思考一位美国女总统可能会具有的特质，即便并不能保证就会有一位女总统。

我们会期待二十六维空间有什么样的性质呢？最明显的一种，即使它**成其为**二十六维的性质，是它通过二十六个数来确定一点，就像两个数确定二维空间中一点、三个数确定三维空间中一点一样。另一个性质是，如果你取一个二十六维形体，使它在各个方向上都扩张两倍，那它的"体积"——假设我们能够使这个词有意义——应当乘以 2^{26}。此外还有其他的一些性质。

若最终发现二十六维空间的概念在逻辑上不能自洽，那这样的思考就没什么意思了。为了消除疑虑，我们终归还是要表明它是存在的——如果它不自洽，明显就不存在——不过是从数学而非物理的角度。这么说的意思是，我们需要定义一个恰当的模型。它可能并不非得是**依附于**某种东西的模型，但如果它具有我们所想要的全部性质，那它就表明这些性质是逻辑自洽的。不过，一如往常，结果表明我们将定义的这个模型非常有用。

怎样定义高维空间？

定义这个模型出奇地容易，只要怀有一个想法——坐标系。我前面说过，二维中的一点可以由两个数确定，三维中则需要三个数。常规做法是采取笛卡尔坐标系，之所以这样称呼是因为它是由笛卡尔发明的（他声称自己是在梦中产生这个想法的）。在二维上，先画出垂直相交的两个方向。例如，一个方向可能向右，另一个方向径直向上，如图21所示。给定平面上任意一点，你都可以通过水平移动一段距离（如果是向左移动，则认为是向右移动了负的距离），再转90度并垂直移动另一距离到达这一点。这两段距离给了你两个数，这两个数就是你所到达的这一点的坐标。图21中显示出了坐标为(3,2)的点（右移三上移二），坐标为(−2,1)的点（左移二上移一）以及坐标为(1,−2)的点（右移一下移二）。完全同样的程序在三维，也就是在空间中

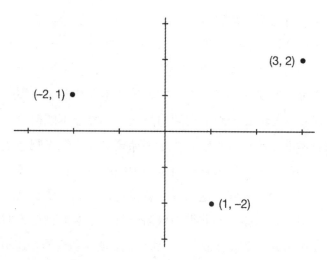

图21 笛卡尔平面上的三点

也有效,只不过你必须使用三个方向,比如向前、向右和向上。

现在让我们稍微改变一下视角。我们不再将这两个数(或三个)称作点在空间中的**坐标**;让我们说,这些数**就是**点。也就是说,我们不再说"坐标为(5,3)的点",让我们说"(5,3)这个点"。有人可能把这仅当作语言上的便利,但它实际上有更深的意义。它在用空间的数学模型取代实在的、物理的空间。我们的二维空间数学模型是由成对的实数(a,b)所组成的。尽管这些成对的数本身并不是空间中的点,我们也称它们为点,因为我们希望提醒自己,这正是它们所表示的东西。类似地,我们可以取所有的实数三元组(a,b,c)得到三维空间的模型,并把这些三元组称为点。现在我们就有了定义诸如八维空间中点的很明显的方法。它们只不过是实数八元组。例如,这就是两个点:$(1,3,-1,4,0,0,6,7)$和$(5,\pi,-3/2,\sqrt{2},17,89.93,-12,\sqrt{2}+1)$。

我现在已经定义了一种初步的数学模型,但它还不值得被称作八维空间的**模型**,因为"空间"一词包含着很多几何含义,我还没有用模型来描述它们:空间不仅仅只是大量单个点的堆砌而已。例如,我们会谈论一对点之间的距离,会谈论直线、圆及其他几何形状。这些思想在高维空间中的对应物是什么呢?

回答这类问题,有一个通用的方法。找出一个二维或三维中熟悉的概念,首先完全用坐标的语言来描述它,然后就能预期它向高维空间中的拓展变得很显然。让我们看看这个方法是怎么处理距离这一概念的。

给定平面上两点,如(1,4)和(5,7),我们可以按如下方法计算它们的距离。首先按图22所示画一个直角三角形,其另一个顶点位于(5,4)。我们注意到,连结(1,4)和(5,7)的线段是这个三角形的斜边,这意味着可以用毕达哥拉斯定理来计算出它的长度。另两条边的长度为5-1=4和7-4=3,所以斜边的长度是$\sqrt{4^2+3^2}=\sqrt{16+9}=5$。因此两点之间的距离为5。将这种方法应用到一般的一对点(a,b)和(c,d)上,我们得到一个直角三角形,其斜边端点正是这两个点,另两条边的长度为$|c-a|$(这表示c和a之间的差距)和$|d-b|$。毕达哥拉斯定理告诉我们,两点间的距离由下式给出:

$$\sqrt{(c-a)^2+(d-b)^2}$$

类似的方法在三维中也有效,只不过稍微复杂一点,可以得出(a,b,c)和(d,e,f)两点间的距离是:

$$\sqrt{(d-a)^2+(e-b)^2+(f-c)^2}$$

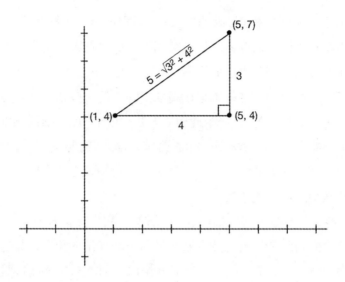

图22　用毕达哥拉斯定理计算距离

换言之，要计算两点间的距离，你需要将对应坐标差的平方相加，然后求平方根。（简要地说，理由如下：以(a,b,c)、(a,b,f)和(d,e,f)为顶点的三角形T，在点(a,b,f)处是个直角。(a,b,c)到(a,b,f)的距离是$|f-c|$，而由二维公式，(a,b,f)到(d,e,f)的距离是$\sqrt{(d-a)^2+(e-b)^2}$。在T内应用毕达哥拉斯定理，就能得到结果。）

这条陈述有个有意思的特征：它没有提到假设点在三维空间中这个事实。因此我们凑巧发现了计算**任意**维空间中距离的方法。例如，$(1,0,-1,4,2)$和$(3,1,1,1,-1)$（五维空间中）这两点间的距离是：

$$\sqrt{(3-1)^2+(1-0)^2+(1-(-1))^2+(1-4)^2+(-1-2)^2} =$$
$$\sqrt{4+1+1+9+9}=\sqrt{27}$$

这种处理方式有一点误导作用，它暗示着任意一对五维空间中的点之间总是有一个距离（要记住，五维空间中的点只不过意味着五个实数而已），而我们发现了怎样把这个距离计算出来。但实际上，我们所做的是**定义**距离的概念。没有什么物理实在强迫我们必须要按上述方法计算五维空间的距离。但另一方面，这种方法很明显，就是我们在二维和三维空间所用方法的自然拓展，若要采用别的定义则会显得很奇怪。

一旦定义了距离，我们就可以开始扩展其他概念。例如，球面显然就是圆在三维空间中的等价物。那四维空间中的"球面"会是什么呢？和距离一样，如果我们把二维和三维版本的概念用一种不提及维数的方法描述出来，就可以回答这个问题了。这其实根本不难：圆和球面都可以通过到定点（圆心或球心）固定距离（半径）的所有点的集合来描述。在此方面，我们完全可以把相同的定义用于四维中的球面，甚至八十七维中的球面。例如，四维空间中以$(1,1,0,0)$为定点以3为半径的球面，正是所有到$(1,1,0,0)$距离为3的（四维的）点所组成的集合。四维空间中的点就是四个实数(a,b,c,d)。它到$(1,1,0,0)$的距离是（根据我们之前的定义）：

$$\sqrt{(a-1)^2+(b-1)^2+c^2+d^2}$$

因此，描述这个四维球面的另一种方式是：它是满足如下条件的所有四元组(a,b,c,d)的集合。

$$\sqrt{(a-1)^2+(b-1)^2+c^2+d^2}=3$$

例如,(1,-1,2,1)就是这样的一个四元组,因此它是这个四维球面上的一点。

另一个可以扩展的概念是二维中的正方形和三维中的立方体。如图23所示,满足 a、b 均在0和1之间的所有点 (a,b) 的集合就形成了边长为1的正方形,它的四个顶点是:$(0,0)$,$(0,1)$,$(1,0)$ 和 $(1,1)$。在三维中,我们可以取满足 a、b、c 均在0和1之间的所有点 (a,b,c) 的集合,以此来定义一个立方体。它的八个顶点是:$(0,0,0)$,$(0,0,1)$,$(0,1,1)$,$(0,1,0)$,$(1,0,0)$,$(1,0,1)$,$(1,1,0)$ 和 $(0,0,0)(1,1,1)$。显然我们可以在高维空间中采用类似的定义。比如,我们可以通过取各个坐标均在0和1之间的所有点 (a,b,c,d,e,f) 的集合,以此得到六维立方体,或者说一个明显当得起这个名字的数学构造。它的顶点就是所有那些坐标非0即1的点。不难看出,维数每增加1,顶点的个数就翻倍,所以这个例子中顶点共有64个。

除了单纯**定义**形状之外,我们还可以做许多其他事情。为了简要说明这一点,我来计算一下五维立方体的边的条数。我们并不能一下子就看出边是什么,但我们可以从二维和三维的情形中得到一些线索:边就是将相邻两个顶点连结起来的线段,两个顶点如果只有一个坐标不同,那就认为它们是相邻的。取五维立方体的一个常见顶点,如 $(0,0,1,0,1)$,根据刚刚给出的定义,它的相邻顶点是 $(1,0,1,0,1)$、$(0,1,1,0,1)$、$(0,0,0,0,1)$、$(0,0,1,1,1)$ 和 $(0,0,1,0,0)$。一般地,每个顶点有五个相邻顶点,因而也就引出五条边。(怎样把连结相邻顶点的线段的概念从二维和三维扩展到五维,我把它留给读者自己思考。对于本次计算,它并不重要。)因为总共有 $2^5=32$ 个顶点,所以好像有 $32\times$

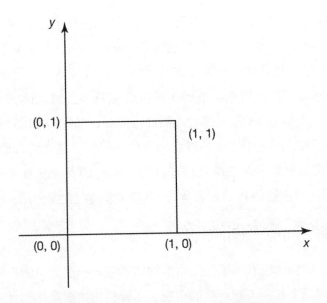

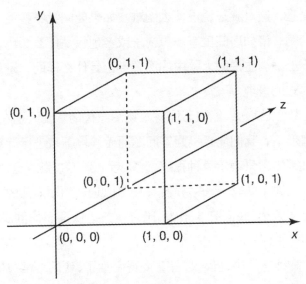

图 23 单位正方形和单位立方体

5=160条边。但是我们把每条边计算了两次——从它的两个端点处各算了一次，所以正确的答案是160的一半，即80条边。

总结上述做法的一种方式是，称之为把几何转化为代数，通过坐标系来把几何概念翻译为等价的但只涉及数之间关系的概念。尽管我们不能直接对几何进行拓展，但我们**可以**对代数进行拓展，而且将这种拓展称为高维几何似乎是合情合理的。五维几何显然不像三维几何那样与我们的切身经验直接相关联，但这并不妨碍我们去思考它，使它作为模型发挥作用。

四维空间能否图像化？

三维物体能够图像化而四维物体不能——这个命题看似显然，实际上却经不起严格的审视。尽管将物体图像化和直接观察它的感觉十分相像，这两种体验却有很重要的区别。例如，如果别人要求我将一个房间画出来，这个房间我比较熟悉但不是十分熟悉，这对我来说没什么困难。如果要问我一些关于它的简单问题，诸如房间里有多少椅子或者地板是什么颜色，我通常答不上来。这说明大脑中的图像不论是什么样的，还是不同于照片式的呈现。

在数学语境中，能与不能将某物图像化的重要区别在于，前一种情形下，我们能直接回答问题而无须停下来进行计算。这个"直接"当然只表示一种程度，但实际上毫不夸张。比如，如果要问我三维立方体有多少条边，我可以通过"仅仅观察"，看到顶面有4条边，底面有4条边，还有4条边从顶面伸到底面，所以一共是12条。

在高维空间中，"仅仅观察"变得困难了，我们常常被迫要像我在讨论五维空间中的推理问题时那样进行更多论证。但观察有时也是可能的。比如，我可以思考一个四维立方体，它由两个彼此相对的三维立方体组成，对应顶点以边（在第四维中）相连，就像三维立方体由两个彼此相对的正方形组成，对应顶点也连结起来一样。尽管我对四维空间没有一个完全清晰的图像，但我仍然可以"看见"两个三维立方体各有 12 条边，8 条边连结着它们对应的顶点。这样，一共就有 12+12+8=32 条边。于是我可以"仅仅观察"到，五维立方体又是由这样的两个四维立方体组成的，依旧是对应顶点相连，总共有 32+32+16=80 条边（每个四维立方体有 32 条边，其间有 16 条边连结它们），恰与我之前得到的答案相同。于是，我具有了某种初步地将四维和五维图像化的能力。（如果你对"图像化"这个词感到困扰，可以换一个词，比如"概念化"。）当然，这远比三维的图像化要困难——比如，我无法直接回答，四维立方体旋转是什么样子，而三维的我就可以说出来——但是，这也明显要比五十三维的图像化要容易，要是它们都不可能的话也就谈不上谁比谁容易的问题了。有一些数学家专攻四维几何，他们四维空间图像化的能力得到了极大拓展。

　　对数学来说，这个心理学要素的影响已远远超出几何学的范围。投身于数学研究所能得到的乐趣之一就是，随着专业领域的经验越来越丰富，你能够发现自己"仅仅观察"就能得到越来越多问题的答案，不一定非得是几何问题，而这些问题你以前可能要艰难思考上一两个小时。举个很基本的例子，来看 471×638=638×471 这条陈述。为了验证，我们可以通过两个很长的竖式乘法来计算，发现它们得到相同的结果。但是，如果考虑一个 471 乘 638 的矩形点阵，你就可以看出，第一个式子是各行点之和，第二个式子是各列点之和，所以它们必然得到相同的结果。注意,在这个问题上,我们头脑中的图像与相片化的图像是很不一样的:你真的看到了一个 471 乘 638 的矩形,而不是 463 乘 641 的矩形吗?你难道能数出短边所有的点来验证吗?

高维几何中的点是什么?

表明能够赋予高维几何某种意义是一回事,但要表明这个问题为什么值得认真对待就是另一回事了。在这一章的前面部分,我曾经说它作为模型是很有用处的。但是,既然我们所居住的实际空间是三维空间,高维几何究竟有什么用处呢?

这个问题的答案相当简单。第一章中我谈到,一个模型可以具有许多不同的功用。即使二维和三维几何也用于许多不同目的,而不仅仅是物理空间的直观模型。例如,我们表示物体的运动时,常常画一张图来记录它所走过的距离随时间的变化。这个图是平面上的一个曲线图,曲线的几何性质与物体运动的信息相对应。为什么二维几何适用于这个运动过程的模型化呢?因为在这里有两个我们关心的数——流逝的时间和走过的距离——如我所说过的,我们可以将二维空间看作所有成对的数的集合。这就提示了我们,为什么高维几何会有用处。宇宙中可能并没有潜藏着高维空间,但需要同时考虑好几个数的情形却有不少。我下面将简要描述两种情形,之后你很明显地可以发现还会有更多的类似情形。

设想我要描述一把椅子的位置。如果是向上直立着的,它的位置就完全是由两条腿与地面接触的点来确定的。这两个点可以分别通过两个坐标来描述。于是四个数就可以用于描述椅子的位置。但这四个数是有联系的,因为椅子腿底端的相互距离是固定的。如果这个距离是 d,地面上两条腿位于点 (p,q) 和

(r,s)，那么由毕达哥拉斯定理有$(p-r)^2+(q-s)^2=d^2$。

这就对p,q,r,s施加了约束，我们可以用几何语言来描述这种约束：四维空间中的点(p,q,r,s)被限制在某个特定的三维"曲面"上。更复杂的物理系统也可以用类似的方式来分析，维度也变得更高。

高维几何在经济学中也很重要。例如，你如果正在犹豫买某个公司的股票是否明智，那么能帮助你进行决策的大多数信息都是以数字的形式出现的——劳动力规模、各种资产的价值、原材料的成本、利率，等等。作为一个序列，这些数可被看作某种高维空间中的一个点。通过分析许多类似的公司，你可能会确定出空间中的某个区域，认为购买此区域中的股票是不错的主意。

分数维

讨论到现在，如果说有什么事情看起来很明显的话，那就是任何形状的维数总是一个整数。你要是说需要两个半坐标来确定一个点——即便是个数学的点，这会是什么意思呢？

这个说法看起来很有道理，但我们遇到的这个困难和在第二章中定义$2^{3/4}$这个数之前遇到的困难非常类似，当时是用抽象方法绕过了它。我们能否对维度做相似的事情呢？要想这么做，我们就必须找到与维度相关的某些性质，它和整数并没有直接的关系。这样就把与坐标数字相关的一切排除在外了，因为坐标看起来和维度这个概念联系实在太紧密了，让人很难思考其他东西。但是，的确还有另一种性质，在本章开头简短地提到过，它恰好给出了我们所需要的东西。

几何有一个重要特征会随维数变化，这就是当把形体沿各方向以因子t扩张时，有个规则决定形体尺寸如何变化。尺寸一

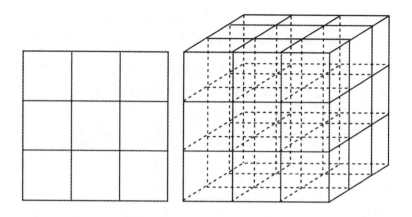

图 24　将正方形分成 $9=3^2$ 个小正方形，将立方体分成 $27=3^3$ 个小立方体

词，我指的是长度、面积或者体积。在一维上尺寸变为 t 倍，或 t^1，在二维上尺寸变为 t^2 倍，在三维上变为 t^3 倍。因此，t 的指数就告诉了我们形体的维数。

到现在为止，我们还没有完全从描述中丢弃整数，因为数字 2 和 3 本身就隐含在"面积"和"体积"这两个词中。我们还可以不用这两个词，而采取如下的办法。为什么边长为 3 的正方形，面积是边长为 1 的正方形的 9 倍？原因是，我们可以把大正方形切成 9 个与小正方形全等的部分（如图 24）。类似地，3 乘 3 乘 3 的立方体，可以分成 27 个 1 乘 1 乘 1 的立方体，所以体积是小立方体的 27 倍。于是，我们可以说，因为当立方体以因子 t 扩张时（其中 t 是大于 1 的整数），则新的立方体可以分成 t^3 个原来的立方体，所以立方体是三维的。注意，"体积"一词没有在上一句话中出现。

现在我们可能会问：像上面一样推理得到的却不是整数，有这样的形体吗？答案是有的。一个最简单的例子就是科赫雪花。我们无法直接描述它，而是要把它定义成下列过程的极限。首先

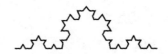

图 25 建造科赫雪花

从一条直线段，比如长度为 1 的线段开始。把线段分成三段相等的部分，以中间一段为底，作一等边三角形，再用三角形的另外两条边来代替中间这一段。结果得到的图形是由四条直线段组成的，每一段的长度都是三分之一。再将每条线段都切成三段相等的部分，仍将所有的中间一段都用等边三角形另两边来替代。现在的结果是 16 条线段，每条线段的长度是九分之一。接下来如何继续这一过程就很清楚了。开始的几步如图 25 所示。不难严格证明，这个过程将导致一个极限形状，正如图中所提示的那样。这个最终形状就是科赫雪花。（取三个这样的形状，连起来形成一个三角形，这样看起来会更像一个雪花。）

科赫雪花有一些很有趣的特征。和我们的主题相关的一个是，这个形状可以用它自身的微小版本建造出来。这仍然可以从图中看出来：它由四个小版组成，每一小版都是由完全版以因子三分之一收缩而得的。现在让我们来考虑，关于维数它告诉了我们什么。

如果一个形体是 d 维的，那当它以因子三分之一收缩时，它

的尺寸会除以 3^d。(如我们所见,当 d 为 1,2,3 时,这是正确的。)这样,如果我们能够用图形的微小版本来构建出原图形,那么我们就需要 3^d 个小版本。因为对于科赫雪花来说需要四个,所以它的维数 d 应当满足 $3^d=4$。由于 $3^1=3$ 而 $3^2=9$,这意味着 d 介于 1 和 2 之间,所以并不是一个整数。实际上,这个数是 $\log_3 4$,约为 1.261 859 5。

这个计算依赖于科赫雪花能够分解为较小的自身这个事实,这是个十分罕见的特征:连圆都不具备这样的特点。不过,我们还可以进一步发展上述思想,给出更为普适的维度定义。正如我们应用抽象方法的其他情况一样,这并不意味着我们发现了科赫雪花以及类似奇异形体的"真正的维度",我们仅仅是找到了与特定性质相容的唯一可能的定义而已。实际上,还有其他定义维度的方法,会对这个问题给出不同的答案。例如,科赫雪花的"拓扑维数"是 1。粗略地讲,这是因为它像直线段一样,删掉内部任何一个点后就分解成为两个不相连的部分。

对于抽象化和一般化这两个孪生的过程,这个例子提供了一点有趣的启示。我已经提到过,要将一个概念一般化,我们应当先找出与其相联系的一些性质,再将这些性质进行一般化。这样做通常只有一种自然的方式,但有时,不同的性质组合会导致不同的一般化,而多种一般化方法有时会硕果累累。

第六章
几何

亘古以来，最具影响力的数学书大概要数约公元前 300 年欧几里得的《几何原本》了。尽管欧几里得生活在两千多年以前，但在多种意义上，他都可以算得上第一位现代数学家——至少是我们所知道的第一位。尤其是，他是第一位系统地使用公理化方法的作者，以五条公理开始全书，并由它们推导出许许多多的几何定理。大多数人熟悉的几何——如果他们的确熟悉某种几何的话，就是欧几里得几何。但从研究的层次来讲，"几何"一词则有着更广泛的定义：当今的几何学家已经不大用到直尺和圆规了。

欧几里得几何

下面是欧几里得的公理。根据通常的惯例，我用"直线"一词表示两端都能无限延伸的线。"直线段"则表示有两个端点的线。

1. 任意两点有且只有一条直线段相连。
2. 任意直线段可以两端延伸形成一条直线，且只能形成一条直线。
3. 给定任意一点 p 及任意长度 r，存在以 r 为半径、p 为圆心的圆。

4. 任意两个直角全等。

5. 直线 N 与两条直线 L 和 M 相交，若 N 的同旁内角之和小于两直角，则 L 和 M 相交于 N 的这一侧。

  第四条和第五条公理如图 26 所示。第四条公理的意思是，如果你将一个直角移动到另一个直角上，它们恰好能够重合。对于第五条公理，由于图中所标的 α 与 β 两角之和小于 180 度，所以我们知道直线 L 和 M 会在 N 右侧的某处相交。第五条公理与所谓的"平行公设"是等价的。平行公设断言，给定任意直线 L 和直线外一点 x，有且只有一条直线 M 经过 x 且永远不与 L 相交。

  欧几里得用这五条公理建造起了全部的几何学，在那之后人们也正是如此去理解几何的。例如，下面就给出了三角形内角和等于 180 度这个著名结果的证明框架。第一步要表明，如果直线 N 与两平行线 L 和 M 相交，则内错角相等。也就是说在类似图 27 的情况下，必定有 $\alpha=\alpha'$ 及 $\beta=\beta'$。这是第五公理的推论。首先，第五公理告诉我们 $\alpha'+\beta$ 至少是 180 度，否则 L 和 M 就会相交（图中直线 N 左侧的某处）。由于 α 和 β 在一起形成一条直线，$\beta=180-\alpha$，有 $\alpha'+(180-\alpha)$ 至少是 180，也就是说 α' 不小于 α。由同样的论证，$\alpha+\beta'=\alpha+(180-\alpha')$ 至少是 180，所以 α 不小于 α'。于是只可能是 α 与 α' 相等。又因为 $\beta=180-\alpha$ 和 $\beta'=180-\alpha'$，于是推出 $\beta=\beta'$。

  现在考虑三角形 ABC，记顶点位于 A、B、C 的三个角分别为 α、β、γ。由第二公理，我们可以将线段 AC 延伸为直线 L。平行公设告诉我们经过 β 有一条直线 M 与 L 不相交。取 α' 和 γ' 为图 28 所示的角。很明显有 $\alpha'+\beta+\gamma'=180$，因为 α'、β 和 γ' 形成一

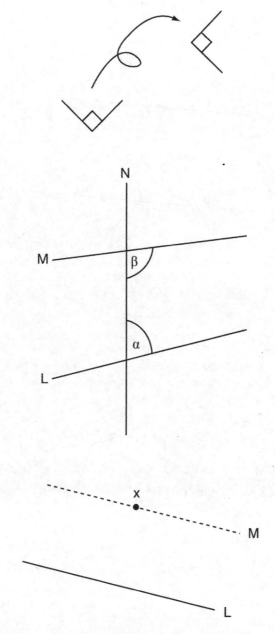

图 26　欧几里得第四公理,及第五公理的两个版本

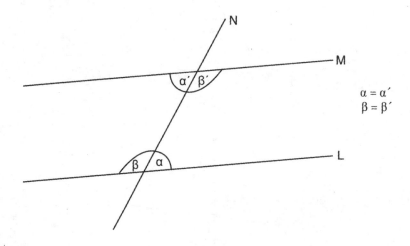

图 27 欧几里得第五公理的推论

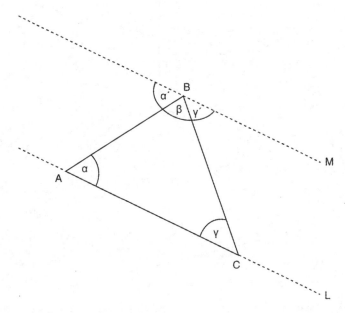

图 28 三角形内角和为 180 度的证明

条直线。因此 $\alpha+\beta+\gamma=180$，正是我们所要求证的结论。

关于日常生活，这个论证告诉了我们什么呢？一个显然的结论似乎是，如果你取空间中 A、B、C 三点，仔细测量三角形 ABC 的三个内角，那它们加起来的和就是 180 度。一个简单的实验就可以证实这一点：在纸上画个三角形，把它尽可能整齐地剪下来，撕成三片，每片包含一个角，将各角拼在一起，可以看到这些角形成了一条直线。

如果你现在确信，不可能想象存在一个物理的三角形，其内角和不等于 180 度，那么你和历史上很多人站在了一起——从公元前 300 年的欧几里得到 18 世纪晚期的伊曼努尔·康德，其间所有人都得到了这个结论。康德非常相信这个结论，以至于他在《纯粹理性批判》中专门用一个显要的部分来谈论怎样绝对确定欧几里得几何为真。

但是康德错了：大约 30 年后，伟大的数学家卡尔·弗雷德里希·高斯就**能够**设想这样一个三角形。有此设想后，他实地去测量了三座山峰构成的三角形的角，分别是汉诺威王国的霍亨哈根山、因瑟尔山和布罗肯山，以检验其内角和是否为 180 度。（这个故事很有名，但诚实的美德迫使我在此注明，人们对于他是否真的想要检验欧几里得几何有一些怀疑。）他的试验不够具有说服力，因为要足够精确地测量这几个角很难，但这个试验的有趣之处与其说在于结果，不如说在于高斯竟然真的去费心尝试。我刚才给出的论证中究竟哪里可能错了呢？

其实，这个问题问得不对，因为上面的**论证**是正确的。但是，既然它以欧几里得五条公理为基础，除非这些公理在日常生活中正确，它就不能提示关于日常生活的任何事情。因此，要怀疑的是论证的**前提条件**，也就是要怀疑欧几里得公理的正确性。

但哪条公理看起来有一点点可疑呢？很难在它们任何一个中找出错误。如果你想在现实世界中用一条直线段连接两个点，那你只需拉紧这条线，使它通过两个点。如果你想把这条直线段延长为直线，那你可以改用激光柱来做到。类似地，画出任意半径和圆心的圆似乎也没有什么难度。经验还告诉我们，拿起纸的两个直角，确实可以使它们恰好重合在一起。最后，有什么能够阻止两条直线像无穷长的铁轨一样一直延伸到无限远？

平行公设

在历史上，引起最多怀疑的——或者至少是最让人感到不放心的公理，就是平行公设。它比其他公理都复杂，并且其基础就涉及到了无穷。当我们证明三角形内角和等于 180 度时，我们必须依赖于空间最远处所发生的事情，这难道不奇怪吗？

让我们更仔细地来检查一下平行公设，并且努力去理解为什么人们觉得它如此显然是正确的。或许在我们的思维深处存在着下列几项论证之一。

(1) 给定直线 L 和直线外一点 x，要通过 x 做一条平行线，只要选经过 x 并和 L **沿着同样的方向**的线即可。

(2) 记 y 为另一个点，它与 x 同在 L 的一侧，且到 L 的距离相同。用直线段连结 x 和 y（公理 1），再将这个直线段延长为直线 M（公理 2）。那么 M 就不会与 L 相交。

(3) 记与 x 在 L 同一侧且到 L 距离相同的点组成的直线为 M。显然它与 L 不相交。

以上这些论证考虑的都是平行于 L 的直线的存在性。下面还有一个更复杂的论证,它想要表明这样的直线至多只有一条,这是平行公设的另一个部分。

(4) 将 L 和 M 用间距相同的平行线段连起来(如图 29 所示,就像铁轨那样),使其中一条线段经过 x。现在假设 N 是另一条经过 x 的直线。在 x 的某一侧,N 必然位于 L 和 M 之间,因此它会与相邻的一条平行线段交于点 u,这个点也位于 L 和 M 之间。假设 u 的位置在 M 到 L 距离的 1% 处,那 N 与下一条线段交点的位置就会在 2% 处,依此类推。于是,经过 100 条线段后,N 将与 L 相交。因为我们对 N 所作的假设仅仅是 N 不同于 M 而已,由此可以得出 M 是经过 x 且不与 L 相交的唯一的直线。

最后,对于经过一点且平行于 L 的直线,这里还有一个貌似

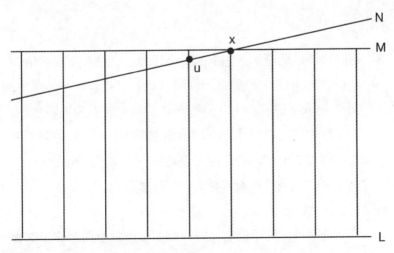

图 29　平行线的唯一性

能同时说明其存在性和唯一性的论证。

(5) 平面上一点可以用笛卡尔坐标系来描述。一条（非竖直）直线 L 有一方程形如 $y=mx+c$。通过改变 c，我们可以上下移动 L。显然，这样得到的直线任意两条都不可能相交，而且平面上任意一点都恰好被一条这样的直线所包含。

注意，我刚才所做的是试图**证明**平行公设，这也恰恰是 19 世纪以前许多数学家试图去做的。他们最希望将平行公设由另外四条公理推导出来，从而表明我们可以忽略掉它。然而，没有人成功过。上面我给出的这些论证及所有类似论证的毛病就在于，它们都隐含了前提假设，当把这些假设明确表达出来时，都不是欧几里得前四条公理的明显推论。尽管这些论证好像都很有道理，但它们并不比平行公设本身**更**有道理。

球面几何

将隐含假设明确表达出来的一个好办法，是在不同的情形下检查同样的论证。怀着这个想法，让我们来看看球面的情况。

要说平行公设在球面上不成立，这并非一望而知地显然，因为球面上根本就没有直线。我们需要使用数学中一种重要的基本思想来绕过这个难题。这种思想是抽象方法发挥作用的一个深刻例子，它就是要**重新解释**直线是什么意思，从而使球面上的确可以包含直线。

其实是有一种自然的定义的：一条线段就是**完全位于球面内**的从 x 到 y 的最短路径。我们可以把 x 和 y 想象为城市，把线

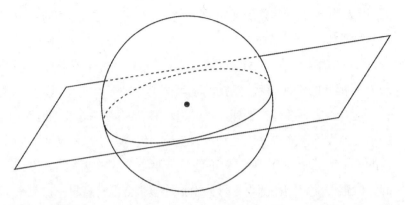

图 30 大圆

段想象为飞机所走的最短路线。这样的路径必定是"大圆",即过球心的平面切割球面所得到的圆(如图 30 所示)的一部分。大圆的一个例子是地球的赤道(为了讨论方便,不妨暂且把地球看作一个严格的球体)。有了定义线段的办法,大圆也就自然成了"直线"的恰当定义。

如果我们采用这样的定义,平行公设当然就不成立了。例

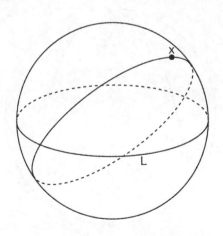

图 31 球面几何中平行公设不成立

如,设 L 为地球的赤道,x 为北半球的一点。不难看出,任何经过 x 的大圆都必有一半在北半球,另一半在南半球,与赤道相交于恰好相对的两点(如图 31 所示)。换句话说,经过 x 不存在一条与 L 不相交的直线(我仍然指的是大圆)。

这看上去好像是在耍花招:如果我用新的办法来定义"直线",那平行公设不再成立并不特别令人吃惊。但令人吃惊不是我们的目的——实际上,我们的定义正是为了要使它不成立才如此设计的。检查一下人们在证明平行公设上的尝试会很有意思。在每个例子中,我们都能够发现在球面几何下无效的隐含假设。

比如,论证(1)假设了,"同样的方向"这个短语的意思很明显。但在球面下这一点完全不明显。为看清这一点,考虑图 32 所示三点,N 为北极点,P 在赤道上,Q 也在赤道上,到 P 的距离为赤道的四分之一。图 32 中还在 P 的位置画了一个小箭头,沿赤道方向指向 Q。那么在 Q 的位置上,指向同样方向的箭头应该

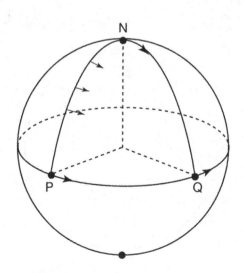

图 32　球面上"同样的方向"没有意义

怎么画？仍然沿着赤道方向是比较自然的。那在 N 处的同方向箭头呢？我们可以这样选择：从 P 到 N 画一条线段。既然 P 点的箭头与这条线段成直角，那 N 点的箭头也应该如此，那实际上也就是应当朝下指向 Q。但是，我们现在就遇到了问题：我们在 N 点画的箭头和在 Q 点画的箭头指的并不是同一个方向。

论证(2)的问题在于不够详细。**为什么**此处定义的直线 M 就不会与 L 相交？毕竟，如果 L 和 M 是球面直线的话，那它们**会**相交。对于论证(3)来说，它假设了 M 是一条直线。对球面来说这不正确：如果 L 是赤道，M 由赤道北边 1000 英里的所有点组成，那么 M 就**不是**大圆。实际上，这是一条纬度恒定的线，所有飞行员和航海员都会告诉你，这不是两点间的最短路线。

论证(4)有点不一样，因为它考虑的是平行线的唯一性而不是存在性。我会在下一小节讨论它。论证(5)作出了一个巨大的假设：这个空间能够用笛卡尔坐标系来描述。但同样，这对球面来说是不正确的。

引入球面几何的意义在于，它让我们可以从论证(1)、(2)、(3)、(5)中分离出某些假设，这些假设实际上是在说："我们所做的几何不是球面几何。"你可能会奇怪，这有什么错呢：毕竟我们做的**不是**球面几何。你可能还会奇怪：如果平行公设确实不是从欧几里得的其他公理中得出，我们怎么才有希望表明这一点呢？说数个世纪以来的数学家努力推导它都以失败而告终是没用的。我们怎么能确定，两百年之内会不会有某位年轻的天才能用绝妙的新思想最终得出证明？

这个问题有个漂亮的回答——至少是在原则上。欧几里得前四条公理的目的在于描述一种有限、平坦、二维空间的几何，但我们并不非要这样去解释它们——至少不是非得按照公理中那种平坦性去解释。如果我们可以将新的含义赋予"直线段"等短语，从而对公理进行重新解释（有人大概会说是"错误解释"），就像我们在球面几何中做的那样；如果我们这样做之后，发现前四条公理都是正确的但平行公设是错误的，那么我们就表明了，平行公设并非从其他公理中推出。

为了看清其中原因，可以想象一种假想的证明，从前四条公理出发，经过一系列严格的逻辑步骤，最终得出平行公设。由于这些步骤都遵循逻辑，如果我们对其赋予新的解释，它们仍会保持有效。但在新的解释下，前四条公理都是正确的，而平行公设不正确，所以这样的论证必然是有错误的。

为什么我们不能正好用球面几何来重新解释呢？原因是，很不幸欧几里得前四条公理在球面上并不全部成立。例如，球面上不能包含半径任意大的圆，所以第三公理不成立；而且从北极到南极不止有一条最短路径，所以第一公理也不成立。所以，尽管球面几何能够帮助我们理解某些尝试过的对平行公设的证明中的缺陷，但它仍然不能保证其他可能成立的证明不存在。因此，我要转向另一种新的解释，称为双曲几何。平行公设在这里再次不成立，但这一次第一到第四公理都是成立的。

双曲几何

描述双曲几何有若干种等价的方式。我所选择的这一种称为圆盘模型，是由伟大的法国数学家亨利·庞加莱所发现的。尽管在这样一本书中我无法确切地给出定义，但我至少能够解释它的一些主要特征，并且讨论一下关于平行公设它能告诉我们什么。

理解圆盘模型比理解球面几何要复杂，因为我们不光要重新解释"直线"和"直线段"等词语,还要重新解释距离这个观念。在球面上,距离有一个很好理解的定义:x和y两点间的距离,就是在球面上从x到y最短路径的长度。尽管双曲几何中类似的定义也是正确的,但并不显然,理由涉及到什么才是最短路径——或者说**任意**路径的长度是什么,我们在下面能看得很清楚。

图33所显示的,是用正五边形对双曲圆盘进行镶嵌。当然我们还需要解释这句话,因为要以通常的方式来理解,图肯定是错误的:显而易见,这些"五边形"的边不是直的,长度也不相同。然而,双曲圆盘中的距离并不是用通常的方式定义的,和常规距离相比,越靠近边缘的距离越大。实际上,虽然看似不像,但边缘处的距离特别大,以至于从圆周上到中心处的距离是无

图33　用正五边形镶嵌双曲平面

穷大。所以，标星号的五边形之所以有一条边看起来比其他四条边都长，原因就是这条边最靠近中心。其他的边可能看起来短，但由双曲距离的定义，这种表面上的短正好由距离边缘的近进行了补偿。

如果这看起来很难解很矛盾，那不妨考虑一幅典型的世界地图。大家都知道，因为地球是圆的而地图是平的，所以距离必然被扭曲了。表示这种扭曲的方式有很多种，最常用的一种是墨卡托投影，靠近极地的国家会显得比它们实际的要大得多。比如格陵兰就看似和整个南美洲大小相当。在这样的地图上，越靠近上端或下端，实际距离就比表面上的距离越小。

这种扭曲会产生一个众所周知的效应，地球表面上两点间最短路线在地图上就显示成了弯的。这种现象可以通过两种途径来理解。第一种是忘掉地图，想象一个地球仪，观察到如果在北半球选两点，第一点在第二点东边很远（巴黎和温哥华是个不错的例子），那么从第一点出发到第二点的最短路径会从靠近北极的地方穿过而不是伸向正西。第二种方法是直接从地图出发，根据越靠近顶部实际距离越短来推理，那么要想缩短旅程，应该同时既向西又向北走。用这种办法很难精确看出最短路径是什么，但至少"直线"（从球面距离的角度看）是弯的（从地图距离的角度看）这条原则是很清楚的。

我前面说过，越靠近双曲圆盘的边缘，与外表距离相比起来，实际距离**越大**。这样的结果是，两点间的最短路径倾向于朝圆盘中心偏折。这也就意味着它不会是通常意义上的直线（除非这条线恰好通过中心）。结果是，双曲直线，即双曲几何观点下的最短路径，正是与大圆边界成直角的圆弧（如图 34 所示）。现在再去观察图 33 的五边形镶嵌图，你会发现五边形的边尽管

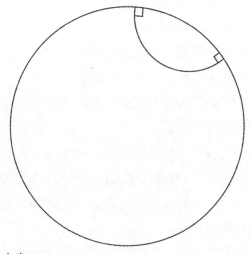

图 34　双曲直线

看起来不是直的,但其实都是双曲直线段,因为根据我刚才给出的定义,它们都可以延伸为双曲直线。类似地,尽管这些五边形的大小和形状看起来不都一样,但实际上的确是一样的,因为靠近边缘处的五边形要比它们看起来的大得多——与格陵兰的例子正相反。如同墨卡托投影一样,圆盘模型也是实际双曲几何的扭曲的"地图"。

　　到这里,有人很自然会问实际的双曲几何是什么样子。也就是说,这个扭曲的地图是反映什么的地图呢?什么东西和圆盘模型的关系,与球面和墨卡托投影的关系是对等的呢?这个问题的确不太好回答。球面几何之所以能够在三维空间中的曲面上实现,这某种意义上是侥幸。如果我们本来是从墨卡托投影**开始**,从它对距离的奇怪概念开始,并不知道这其实是球面的地图,那么我们要是发现它恰好对应于一个完美对称的曲面,正是将曲面投射到平面的地图,而地图距离只不过是常规意义下容易理解的曲面最短路径长度,那我们一定会既惊讶又高兴。

不幸的是,对于双曲几何来说这种情况却不存在。但有意思的是,这并没有使双曲几何比起球面几何来缺乏实在性。这有点难以理解——至少刚开始有点,但正如我在第二章中强调的,数学概念的实在性更多地与它做什么而不是与它是什么相关。因为我们能够说清双曲圆盘做什么(例如,你要问我,将五边形镶嵌图沿中心五边形的一个顶点旋转 30 度是什么意思,那我是能够回答你的),所以双曲几何就和其他所有数学概念一样真实。从三维欧氏几何的视角来看,球面几何可能更易于理解,但这并不构成根本的差异。

双曲几何的另一个性质是它满足欧几里得前四条公理。例如,任意两点都可以且仅可以被一条双曲直线段(即与主圆垂直相交的圆弧)相连。尽管看上去好像不能以任意圆心画半径比较长的圆,但这是因为你忘了,越靠近圆盘边缘距离越大。实

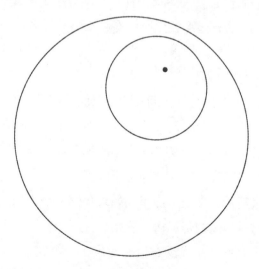

图 35 双曲圆及圆心

际上，如果双曲圆几乎要擦上圆盘边缘了，那它的半径（双曲半径）将会非常大。（双曲圆恰好与普通圆外形一样，但它们的圆心却不在我们所料想的位置。参见图35。）

至于平行公设，如我们所期望的，它对于双曲几何不成立。这可以从图36中看出，在图中我将三条（双曲）直线标为 L、M_1 和 M_2。直线 M_1 和 M_2 在 x 点相交，但它们都不与 L 相交。因此，经过 x 有两条（实际上有无穷多）直线不与 L 相交。这违背了平行公设，平行公设说的是只能有一条直线。换言之，我们在双曲几何中恰好如愿找到了对欧几里得公理新的解释，从而说明了平行公设不是另外四条公理的推论。

当然，在本书中我其实没有证明双曲几何具有我所声称的全部性质。要这样做的话，需要在一般的大学数学课程中花上好几次课的时间，不过至少我还能就如何定义双曲距离说得更确切一些。要说清它，我就必须明确说明，在靠近圆盘边缘处，

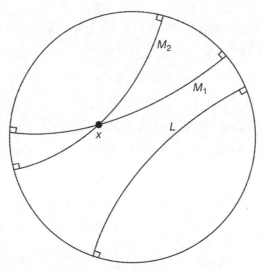

图36　平行公设在双曲平面中不成立

实际距离比看起来大了**多少**。答案是,点 P 处的双曲距离是"常规"距离的 $1/d^2$ 倍,其中 d 是 P 到主圆边界的(常规)距离。换一种说法,如果你在双曲圆盘中移动,那么根据双曲距离的概念,你经过 P 点的速度是你表面速度的 $1/d^2$ 倍。这就意味着,如果你保持恒定的双曲速度,那当你接近圆盘边界时,看起来速度就会越来越慢。

离开双曲几何之前,让我们来看看我前面给出的论证(4)为何不能证明平行线的唯一性。它的想法是这样的:给定直线 L 和直线外一点 x,过 x 画一条直线 M 与 L 不相交,我们可以用几条垂直于 L 和 M 的线段将两条线连起来,将 L 和 M 之间的空间划分为矩形。这件事看似显然能够做到,但在双曲世界中却是不可能的,因为在其中四边形的内角和总是**小于**360度。换句话说,在双曲圆盘中,论证所需的矩形是不存在的。

空间何以能够弯曲?

数学中(以及物理学中)听起来最自相矛盾的短语之一就是"弯曲空间"。我们都知道线或面被弯曲是什么意思,但空间本身就是**自在之物**。即便我们能够在一定程度上对三维空间弯曲的概念赋予意义,与曲面的类比还是揭示出,我们自己不可能观察到空间是否弯曲,除非跳到第四维中去观察。在那里也许我们会发现宇宙是一个四维球体(我在第五章中解释过的概念)的表面,这个球面至少听起来是弯曲的。

当然,这些都是不可能的。因为我们不知道如何能站到宇宙之外——这种想法几乎在措辞上就是矛盾的——我们能够用的证据只能来自于宇宙之内。那么,什么样的证据有可能说服我们空间是弯曲的呢?

和之前一样,如果我们采取抽象方法,这个问题就变得简单些了。不去做艰深的思维体操以试图理解弯曲空间的本性,让我们仅仅遵循扩展数学概念的寻常程序。我们理解"弯曲"一词用在二维表面时的意思。为了把它用在不太熟悉的情形中,即用到三维上,我们必须努力找到弯曲表面易于扩展的**性质**,就像我们要定义 $2^{3/2}$、五维立方体或科赫雪花的维度时所做的那样。因为我们最终要找到的性质应当是能够在空间**之内**察觉到的,所以我们应当去寻找,怎样在不需要站到弯曲表面之外的情况下就能察觉到它的弯曲性。

比如,确信地球表面是弯曲的一种办法是乘坐航天飞机向地面看,发现它近似球形。但是下面的这个更接近二维情形的试验,同样非常有说服力。从北极点出发一直向正南走大约6200英里,记下你初始的方向。然后向左转,再走相同的距离。然后再向左转,再走一次相同的距离。6200英里大致是北极点到赤道的距离,所以你的旅程会把你从北极点带到赤道,绕赤道走过四分之一,然后再次回到北极点。而且,你回到北极点时的方向应当与你出发的方向夹一直角。于是我们得到,在地球表面上,有一个各角均等于直角的等边三角形。在平坦表面上,等边三角形的内角必须相等且和为180度,所以各角均为60度。因此地球表面不是平坦的。

于是,从曲面内部说明二维曲面弯曲的一种方法,就是找出内角和不为180度的三角形,而且这种办法也是可以在三维中尝试的。这一章我主要关注二维中的欧氏几何、球面几何和双曲几何,但它们都很容易扩展到三维上去。如果我们测量空间三角形的角度,发现它们的和大于180度,这就说明空间更接近于三维版本的球体表面,而不太接近于能用三个笛卡尔坐标描述的那类空间。

如果事实确实如此，那么说空间是正性弯曲的似乎就很合理了。这个空间可以预料到的另一个性质是，沿同一方向出发的直线都会相互靠近并最终相交。还有一个性质，半径为r的圆的周长不是$2\pi r$，要比这个数小一些。

你可能很想指出，我们所知的这个空间并没有这些特性。沿相同方向的直线一直都能保持相同的方向，三角形内角和及圆的周长都如它们所应是的那样。换句话说，即便空间在逻辑上是可能弯曲的，它看起来其实也是平坦的。但是，空间对我们而言看上去平坦，有可能只是因为我们生活在空间特别小的一部分之中。这就像对那些没走过太远的人来说，地球表面看起来也是平的——至少除了大大小小的凹凸外基本平坦。

换句话说，空间可能只是**粗略地**平坦。如果我们能够画一个非常大的三角形，或许就会发现它的内角和不是180度。这就是高斯试图做的，可他的三角形却根本不够大。然而，在1919年，史上最著名的科学实验之一显示，弯曲空间的思想不仅是数学家的迷思，更是生活中的事实。根据爱因斯坦四年前发表的广义相对论，空间因引力而弯曲，因而光的行进并不总是沿直线——至少是按欧几里得的意思来理解的直线。这个效应很微弱，很难轻易探测到，但1919年一次日全食提供了很好的机会。这次日全食可以在几内亚湾的普林西比岛观测到。日食发生时，物理学家阿瑟·爱丁顿拍摄下了照片，照片显示靠近太阳的星星并不在通常预期的位置上，这正符合爱因斯坦理论的预测。

尽管人们现在已经接受空间（更精确地讲，是时空）是弯曲的，但也有可能正像地球表面的山峦和谷地一样，我们所观测到的曲率只不过是某个更为庞大、更为对称的形状上的小摄

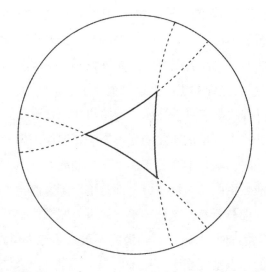

图37 双曲三角形

动。天文学中一个重大的未决问题就是去确定宇宙的**大尺度**形状,即将恒星、黑洞等造成的弯曲熨平后宇宙的形状。它是仍然像大球面一样是弯曲的呢,还是像我们自然而然却很可能错误地想象的那样,是平坦的呢?

还有第三种可能性,宇宙是**负性**弯曲的。容易理解,这是指或多或少与正性弯曲相反。所以,负性弯曲的证据就会是,三角形内角和**小于180度**,沿相同方向的直线会发散,或者是半径为 r 的圆的周长**大于** $2\pi r$。这类行为在双曲圆盘上会发生。例如,图37显示了内角和明显小于180度的三角形。要将球面和双曲圆盘扩展到高维的类似情形并不难,而且就大尺度的时空形状而言,双曲几何有可能是比球面几何和欧氏几何更好的模型。

流形

闭曲面是指没有边界的二维形状。球面就是个不错的例子,环面(即铁圈或面包圈的形状的数学称谓)也是。对曲率的讨论表明,曲面虽存在于三维空间之中,但若脱离开三维空间的参照来思考曲面,可能会很有益。如果我们要将闭曲面的概念扩展到高维空间,这样的想法就更加重要了。

不是只有数学家愿意从纯二维的角度来思考曲面。比如,美国的几何结构受到地球弯曲的显著影响,但如果想设计一幅实用的公路地图,则并不需要印在单张弯曲的大纸上。更实际的办法是印成数页的书,每一页上都是这个国家的一小部分。这些局部最好部分重叠,这样一来,如果有城镇位于某一页的边缘,显得不太方便,那它在另一页将不再处于边缘位置。另外,在每一页的边上都要指明,哪一页包含了重叠的区域,重叠以怎样的形式出现。由于地球的弯曲,没有哪一页是绝对精确的,但我们可以在图中画出经线和纬线以指明微小的扭曲。用这种办法,我们就可以用书中平坦的几页纸把美国的几何结构完全包纳了。

原则上,我们可以用类似方法制出精度相当的世界地图册(不过会有很多页几乎全是蓝色的)。因此,球面的数学性质就能以这种方式被一本地图册所涵盖。如果你想要回答有关球面的几何问题,却完全没有能力将球面图像化,只要你手头上有一本地图册,稍加努力就能做得到。图38显示了一个九页的地图册,画的并不是球面,而是环面。要看出它是如何对应于环形曲面的,可以想象一下,将各页粘在一起形成一大页,然后将大页的上下粘合形成圆柱,最后再将圆柱两端粘接。

数学中一大重要分支研究的就是称为流形的对象。流形正是将上述思想扩展到三维或更高维所得的结果。粗略地讲,一个d维流形就是任何一个这样的几何对象,其中任意一点都会

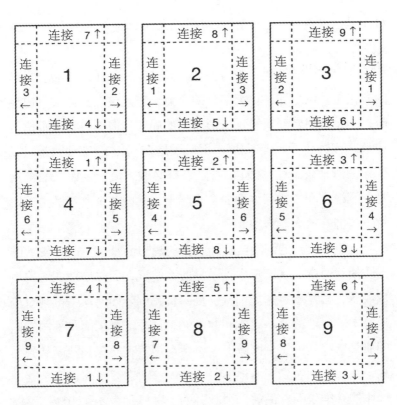

图38　环面地图册

被一小块极为类似于 d 维空间的区域所包围。由于随着维数的增加，流形会变得越来越难以图像化，所以地图册的思想也就相应变得更加有用了。

让我们稍加考虑一下，一个三维流形的地图册会是什么样的。其中每一页当然都必须是三维的，而且和公路图的页面一样应当是平坦的。所谓平坦，也就是应当为一块块我们熟悉的欧氏空间。我们可以要求它们都是长方体，但这在数学上并不太重要。这些三维"图页"中的每一页都是流形上一小部分的地图，我们会仔细地标记出各页之间如何重叠。一种典型的标记可能是这样的：在A页某边缘处的(x,y,z)点对应于B页上的$(2y, x, z-4)$点。

给定了这样的地图册，要如何想象在流形上的移动呢？显然的方式是，设想在某页上移动的一点，一旦该点到达该页边缘时，还会有另外一页表示流形相同的局部，但点在其中不位于边缘，所以我们可以转向那一页。于是，流形的全部几何都可以用地图册来公式化地表达，所以没有必要将流形视作"确实"置身于四维空间中的三维表面。实际上，某些三维流形甚至根本就不可能被放入四维空间中。

一些很自然的问题会随地图册的思想而来。例如，尽管地图册能使我们说明在流形中移动是什么情况，但如果地图册的页数太多，相互重叠的规则特别复杂，我们要如何从中得到对流形基本"形状"的一些感觉呢？我们怎样才能辨别两本不同的地图册其实都是关于相同流形的？特别是，有没有某种简易的办法，让我们一看到三维地图册，就能辨别它表示的是不是四维球体的三维表面？最后一个问题的一种精确表述称为庞加莱猜想，尚未解决；解答这一问题将能够得到100万美元的奖金（由克雷数学研究所提供）[1]。

---

[1] 庞加莱猜想现已由俄罗斯数学家佩雷尔曼最终解决。

第七章
估计与近似

大多数人认为数学是一门纯净、精确的学科。我们经过在中小学的学习,料想数学问题如果可以被简洁地陈述,大概就能得到简练地回答,通常是一个简单的公式。而继续学习大学阶段数学的人,尤其是那些专门研究数学的人,很快就发现这样的想法实在是大错特错。对于很多问题来说,如果有人能够找到解答的精确公式,那简直完全出人意料,如同奇迹一般。多数情况下,我们不得不满足于大致的估计。在你对此感到习以为常之前,这些估计总是看似很丑陋,难以令人满意。然而,品尝一下其中的滋味也是值得的,否则你就会错过数学中很多最伟大的定理以及最有趣的未解决问题。

无法用简单公式表达的简单序列

记 $a_1, a_2, a_3, \cdots$ 为一列实数,根据如下规则产生。第一个数 a_1 等于 1,后面各个数都是前一个数加上其平方根。也就是,对任意 n,我们让 $a_{n+1} = a_n + \sqrt{a_n}$。基于这个简单的规则,显然可以提出问题:$a_n$ 是什么数?

为了对这个问题有些感性认识,让我们对较小的几个 n 来算

出 a_n。我们有 $a_2=1+\sqrt{1}=2$。接下来:

$$a_3=a_2+\sqrt{a_2}=2+\sqrt{2}$$

$$a_4=a_3+\sqrt{a_3}=2+\sqrt{2}+\sqrt{2+\sqrt{2}}$$

$$a_5=a_4+\sqrt{a_4}=2+\sqrt{2}+\sqrt{2+\sqrt{2}}+\sqrt{2+\sqrt{2}+\sqrt{2+\sqrt{2}}}$$

等等等等。注意,等号右端的表达式看起来无法简化,而且每个新的表达式都是前一个长度的两倍。从这个观察中,很容易得出,数字 a_{12} 的表达式中将会出现1024个2,它们中的大多数都深深地位于根号丛中。这样的表达式并不会使我们对数字 a_{12} 有多深的了解。

那么我们因此就应当放弃去理解这个序列吗?不。尽管当 n 比较大时,要考虑 a_n 的精确值似乎没有好办法,但这并未排除得到较准确估计的可能性。实际上,一个好的估计最终可能反而会更有用处。我上面写出了 a_5 的精确表达式,但和 a_5 约为 $7\frac{1}{2}$ 这条信息比起来,这个表达式让你更好地理解 a_5 了吗?

所以,让我们不要再问 a_n 是什么,转而去问 a_n 大约是多大。也就是,让我们寻找能给出 a_n 较优近似值的简单公式。这样的公式是存在的: a_n 大约为 $n^2/4$。要严格地证明需要一点技巧,但要看出为什么这个估计是合理的,我们要注意:

$$\frac{(n+1)^2}{4}=\frac{n^2+2n+1}{4}=\frac{n^2}{4}+\frac{n}{2}+\frac{1}{4}=\frac{n^2}{4}+\sqrt{\frac{n^2}{4}}+\frac{1}{4}$$

也就是说，如果$b_n=n^2/4$，那么$b_{n+1}=b_n+\sqrt{b_n}+1/4$。要是没有其中的"$+1/4$"的话，这个式子就会告诉我们$b_n$正和$a_n$的生成方法一样。而当$n$较大时，加上$1/4$"只不过是个小扰动"（这就是我略去证明的部分），所以b_n可以看作近似正确地生成，由此推出b_n即$n^2/4$给出了a_n较好的近似，正如我之前所断言的。

近似的方法

作类似于刚才的关于近似的论断时，明确什么才算是较好的近似很重要，因为标准会随着情景的不同而变化。如果想用一条能够简单定义的序列$b_1,b_2,b_3,\cdots$来近似一条稳定增大的序列$a_1,a_2,a_3,\cdots$，那我们所能期待的最优近似——实则很少能够达到，就是每一对a_n和b_n的差距都小于一定值——诸如1000。那么随着a_n和b_n增大，它们的**比值**会非常接近于1。例如，假设某时有$a_n=2408597348632498758828$，以及$b_n=2408597348632498759734$。那么$b_n-a_n=906$，这个差虽然是个大数，但与$a_n$和$b_n$比起来就微不足道了。如果$b_n$是在这种意义上对$a_n$的近似，我们可以说$a_n$和$b_n$是"相差常数以内的相等"。

另一类较优的近似是，随着n的增大，a_n和b_n的比值变得非常接近1。当a_n和b_n相差常数以内相等时，这种情况是成立的，但它在其他一些情况下也会成立。例如，如果$a_n=n^2$而$b_n=n^2+3n$，则比值b_n/a_n为$1+3/n$，在n较大时很接近1，尽管a_n和b_n的**差**$3n$会很大。

即便这种情况常常也是难以企及的，如果能找到更弱的近似方法我们就会很高兴了。常见的一种方法是，如果a_n和b_n"相差常数**倍**以内相等"，就视之为近似相等。它的意思是，a_n/b_n和b_n/a_n都不会超过某个固定的数——诸如1000也是可能的，但越小越好。换言之，此时不是 a_n 和 b_n 的差，而是它们的比值保持在某个限度内。

要说两个相差1000倍的数大致相同，这或许看似很不合情理。但这是因为我们总习惯于处理较小的数。自然没有人会将17和13 895看作大致相同，但宽泛地讲，要说下面这样两个数规模**相类**则并不显得很荒唐：

290475629408976156238954534598760889079687234751434875775468 和
360982345987209761234987098249863408762345779678458734598716 6464

尽管第二个数比第一个数大1000倍以上，但它们的数位大体相同——都在60到65位之间。既然没有其他有趣的性质，这一点很可能就是我们所关注的全部。

如果连**这样**程度的近似都是奢求，那也常常值得去找出两条参考序列 $b_1, b_2, b_3, \cdots$ 和 $c_1, c_2, c_3, \cdots$，并证明 b_n 总小于 a_n，而 c_n 总大于 a_n。那么我们可以说 b_n 是 a_n 的"下界"，c_n 是 a_n 的"上界"。比如，一位数学家想估计某个量 a_n，他可能会说："我不知道 a_n 的值是多少，甚至连近似值也找不到，但我能够证明它至少是 $n^3/2$，但不大于 n^3。"如果这个问题相当困难，那这样的一个定理就可能成为重大进展。

关于对数、平方根等你只需要知道这些

估计和近似遍布于数学之中，而这个领域以外的人却不太了解这一点，部分原因在于，为了谈论近似，要使用类似于"大约和 $\log n$ 的速度一样快"，或者"限制在相差一个常数内为根号 t"这样的语言，而这对多数人来说意义不大。幸而，对于较大数

的对数或平方根，如果有人只关心**近似**值，便很容易理解，这一类语言也不难懂。

如果你取两个大的正整数m和n，想快速得出乘积mn的估值，那你应该怎么做？较好的办法是，一开始先数一下m和n各自的位数，如果m有h位，n有k位，那么m位于10^{h-1}和10^h之间，n位于10^{k-1}和10^k之间，所以mn位于10^{h+k-2}和10^{h+k}之间。于是，仅仅数一数m和n的位数就可以"在100倍以内"决定mn，即mn必落在10^{h+k-2}和10^{h+k}这两个数之间，且10^{h+k}仅比10^{h+k-2}大100倍。如果你折中一下，取10^{h+k-1}作为估计值，那么它与mn相差的因子至多为10。

换句话说，如果你只在"相差常数倍以内"的程度上关心某个数，那么乘法立刻就变简单了：取m和n两个数，数数它们合起来的位数，减去1（如果你在乎这一点），写下一个有这么多位的数。例如，1 293 875（7位）乘以20 986 759 777（11位）大致接近10 000 000 000 000 000（17位）。如果你想做得更仔细一点，那么可以用上第一个数的首位1及第二个数的首位2，也就意味着20 000 000 000 000 000是个更优的估计值，不过在很多场合下这样的精度是不必要的。

既然近似乘法这么简单，近似平方也就很简单了——只要把原来的数替换为一个两倍位数的新数。由此得出，把n的位数**减半**可以求出n平方根的近似值。类似地，将位数除以3可以求出立方根的近似值。更一般地，如果n是个大整数，t是任一正数，那么n^t的位数大约为n的位数乘以t。

关于对数呢？从近似的观点看它们其实极为简单：一个数的对数基本上就是它所包含的位数。例如，34 587和492 348 797 548 735的对数分别约为5和15。

实际上，数一个数的位数所近似的是它以10为底的对数，即在便携计算器上按LOG键得到的数。通常,数学家谈论对数指的都是所谓的"自然"对数,也就是以e为底的对数。尽管e这个数的确很自然也很重要,但我们在此只要了解:一个数的自然对数,即在计算器上按LN键得到的数,大体上是它的位数乘以2.3上下。于是2 305 799 985 748的自然对数约为:$13 \times 2.3 = 29.9$。(如果你了解对数,你会知道真正应当乘上的数是$\log_e 10$。)

这一过程也可以反过来进行。假设你有一个数t且知道它是另一个数n的自然对数。那么数n称作t的**指数幂**,记作e^t。那么n大致是多少呢?刚才为了得到t,我们数n的位数并乘以2.3。所以n的位数必定约为$t/2.3$。这就决定了n的大小,至少是近似值。

我上面给出的近似值定义最主要的用途是作比较。例如,现在我们能够明显看出,对大数n而言,其对数要远小于立方根:比如,如果n有75位,它的立方根将会很大——大约有25位数,但它的自然对数则大约仅为$75 \times 2.3 = 172.5$。类似地,数m的指数幂会远大于它的乘方,如m^{10}。例如,若m有50位,那么m^{10}大约有500位,但e^m大约有$m/2.3$位,远大于500。

对数字$n=941\ 192$施行不同的运算,得到的近似结果如下表所示。我没有包含进e^n,因为如果要这么做,那我就得给这本书换个名字了。

n	941192
n^2	885842380864
$\sqrt{n}$	970.15
$\sqrt[3]{n}$	98
$\log_e n$	13.755

$$\log_{10} n \qquad 5.974$$

素数定理

素数是大于1且不能被其他整数——1和自身显然除外——整除的整数。小于150的所有素数有2, 3, 5, 7, 11, 13, 17, 19, 23, 29, 31, 37, 41, 43, 47, 53, 59, 61, 67, 71, 73, 79, 83, 89, 97, 101, 103, 107, 109, 113, 127, 131, 137, 139和149。小于150的其他数都可以做因子分解：例如$91=7 \times 13$。（你可能会担心，将1排除在素数的定义之外似乎不合情理。这并不表达数的某种深层事实，只是碰巧成为一个有用的惯例，采纳这样的定义能使任意数被分解为素数的方式仅有一种。）

自从古希腊时期以来，素数就一直困扰着数学家们，因为它们表面上多多少少是随机分布的，但又并非全然随机。从没有人找出一种简单的规则，能够告诉我们第n个素数是多少（当然可以下番工夫列出前n个素数，但这算不上是简单的规则，而且当n很大时会很不切实际），几乎也不太可能会有这么一种规则。但另一方面，仅仅检查前35个素数，也能向我们透露一些有意思的特征。如果你算出相邻两个素数的差，那么你会得到下面这个新的序列：1, 2, 2, 4, 2, 4, 2, 4, 6, 2, 6, 4, 2, 4, 6, 6, 2, 6, 4, 2, 6, 4, 6, 8, 4, 2, 4, 2, 4, 14, 4, 6, 2, 10。（即$1=3-2, 2=5-3, 2=7-5, 4=11-7$等等。）这列数仍显得有些不规则，但这些数有一个趋势，大致可以看出它们逐渐在增大。显然它们不是稳定地增大的，但最开始的几个数都不大于4，而10和14这样的数直到后来才会出现。

如果你写出前1000个素数，那么相邻素数间距增大的趋势就会变得更加明显。也就是说，和小素数比起来，大素数的出现越来越稀疏。我们正可以预料到这一点，因为大数**不成为**素数的方式更多了。例如，可能会有人猜测10 001是素数，尤其因为它无法被2，3，5，7，11，13，17以及19整除——但实际上它等于73×137。

有自尊心的数学家绝不会满足于仅仅观察到（甚至没有得到严格地证明）大素数比小素数稀少。他一定想要知道，它们稀少到**何种具体程度**。如果你在1 000 001和1 010 000之间随机取一数，那么这个数有多大的机会是素数？换言之，1 000 000附近的素数"密度"是多大？它是极其小还是仅仅比较小？

没有接触过大学数学的人很少会提出这样的问题，其原因在于，他们没掌握将问题公式化表达并进一步思考所需的语言。不过，若是这章到现在为止你都看懂了，那么你就能够欣赏到数学中最伟大的成就之一：素数定理。定理陈述的是，在数 n 附近的素数密度约为 $1/\log_e n$，即1除以 n 的自然对数。

现在再来考虑1 000 001和1 010 000之间取随机数为素数的机会大小。这个区间内的数都大体等于100万。按照素数定理，密度因而约为1除以100万的自然对数。它以10为底的对数是6（在这个例子中，数位数会得到7，但既然我们知道精确答案，不妨使用精确值），所以自然对数约为6×2.3，即13.8。因此，1 000 001和1 010 000之间平均每14个数中有1个是素数，即略多于7%。相比之下，小于100的素数共有24个，即几乎占所有数的四分之一，这就说明了随着数的增大，素数密度是如何减小的。

既然素数分布有零零散散、颇似随机的性质，而我们却能证明其如此多的特点，这足以令人十分惊讶。有意思的是，关于素数的定理通常都是通过**利用**这种看似随机的性质得到证明的。例如，维诺格拉多夫在1937年证明的一个著名定理认为，任意充分大的奇数都可以分解为三个素数之和。我无法在本书中解释他是怎样证明的，但他绝对没有找出将奇数表达为三素数之和的**方法**。这样的思路几乎注定会失败，因为即使是生成这些素数本身也非常困难。基于哈代和利特伍德之前的工作，他大体按照下述办法来论证。如果你能够按照和素数分布同样的密度来真正随机地选取一些数，那么概率论的某种初步理论就能够表明，你几乎一定能够将所有充分大的数表示为你所取的这些数中的三个之和。实际上，你能够以多种不同方式进行这一分解。因为素数是类似于随机的（证明中较难的部分就是要说明，"类似于随机"是什么意思，再加以严格证明），它们的行为就相仿于随机选取的序列，所以所有充分大的数都是三素数之和——同样也以多种方式。为了解释这种现象，这里我们以35为例，列出它分解为三素数之和的所有方式：

$$35=2+2+31=3+3+29=3+13+19=5+7+23$$
$$=5+11+19=5+13+17=7+11+17=11+11+13$$

关于素数的很多研究都具有此类特点。你首先对素数设计一种概率模型，即假装告诉自己，它们是根据某种随机过程挑选出来的。接下来，在假设素数的确是随机产生的情况下，求证有哪些论断是正确的。这样可以使你猜测出很多问题的答案。最后，你努力表明，这个模型足够现实，能够保证你的猜测近似准确。要注意的是，如果强迫在论证中的每一步都给出精确答案，那这个思路就是不可能的。

很有意思，概率模型不仅仅是物理现象的模型，还能成为另一数学分支的模型。尽管素数的真实分布是严格确定下来的，可某种程度上它们看起来也像是实验数据。一旦这样看待它们，我们就很想去设计对应的简化模型，来预测特定概率论问题的答案是什么样的。这种模型有时的确曾使人们得到对素数本身的有效证明。

这种风格的论证取得了某些显著的成功，但它依然没能解决许多著名问题。例如，哥德巴赫猜想断言，任意大于4的偶数都可以表示为两个奇素数之和。这个猜想看起来比维诺格拉多夫所解答的三素数猜想要难得多。此外还有孪生素数猜想[1]，它声称有无穷对相距为2的素数，诸如17与19，137与139。表达这个问题的另一种方式是，如果你写出相邻素数之差，就像我之前做过的那样，那么2永远都会出现（尽管越来越稀少）。

数学中最著名的未决问题大概要数黎曼假设了。这个假设有若干种等价的表达形式，其中一种涉及素数定理给出的估计的精度。我之前说过，素数定理告诉我们在某数附近素数的近似密度。根据这个信息，我们可以计算出，不大于n的素数大约有多少。但这个"大约"有多"大约"？如果$p(n)$是不大于n的素数个数的真实值，$q(n)$是根据素数定理得到的估计值，那么黎曼猜想断言，$p(n)$和$q(n)$的差不会比$\sqrt{n}$大太多。如果能够证明这样的精度确实成立，那么它将会有很多的应用，但迄今所得到的结果比这要弱得多。

排序算法

数学另有一个分支领域，理论计算机科学，其中满是**粗略的估计**。如果有人要写一个计算机程序完成特定任务，那么程序

---

[1] 2013年，华人数学家张益唐证明了弱形式的孪生素数猜想。

当然运行得越快越好。理论计算机科学家提出了这样的问题：我们所能期望的最快速度是多快？

要精确地回答这个问题几乎总是不现实的，所以我们转而证明诸如"当输入规模为n时，这种算法的运行步数约为n^2"这样的论断。由此出发我们可以总结出，一台普通的个人电脑有可能处理的输入规模（简言之，即你想要分析的信息量）为1000，而不是1 000 000。于是，这样的估计就有了实际的重要性。

计算机能够完成的一项非常有用的任务被称为排序，即将为数很多的对象按照给定标准排列顺序。举例来说，想象一下你想要根据喜好程度来排列一系列对象（不一定非要是无生命的，也有可能是一项工作的应聘者之类）。设想，你无法对每个对象赋予一数值来表示有多喜欢它，但是若给定任意两个对象，你总能确定更喜欢哪一个。再设想，你的偏好是前后一致的，意思是指，永远不可能出现这样的情况：喜欢A胜于B，喜欢B胜于C，又喜欢C胜于A。如果你不想为这个任务花太多时间，那么将比较次数最小化就会很有意义。

当对象很少时，很容易看出要怎样做。比如有两个对象，那你至少要作一次比较，一旦比较了你也就知道了它们的排列顺序。若有三个对象A、B、C，那么一次比较是不够的，但总得从某个比较开始，从哪个开始都无妨。为了论证，假设你先比较了A和B，更喜欢A。这时你必须用A和B中的一个和C进行比较。如果你用A和C来比较，并且更喜欢C，那么你就知道了你的偏好顺序是C，A，B；但如果你发现更喜欢A，那么关于B和C，你就只知道它们都没有A讨你喜欢。所以这时需要进行第三次比较，以便能将B和C加入排序。综上所述，三次比较总是充分的，而且有时也是必要的。

有四个对象A、B、C、D时情况如何呢？分析起来就更加困难了。你可能仍从比较A和B开始。但一旦比完这一对，接下来的比较就会有两种本质不同的可能性。一种是用A或B来比较C，另一种是比较C和D，哪种主意更好并不明显。

设想你要比较B和C。如果幸运，你就能够对A、B、C进行排序。假设排序就是A，B，C。接下来还需要考虑D适合什么位置。最好的办法是先比D和B。在这之后，你就只需再比较一次，即比较D和A（如果D比B好）或比较D和C（如果B比D好）。以上一共比较了四次——两次来确定A、B、C的顺序，两次找出D的位置。

我们还没有分析完这个问题，因为在A、B、C的比较中你可能并不走运。也许比完前两次以后，你只知道A和C都比B好。那么面临着另一个困境：是比较A和C更优呢，还是用D来与A、B和C中的一个比较更优呢？考虑完各种情况及各种子情况，你还需要再观察，如果第二次比较的是C和D，情况会如何。

分析变得烦琐起来，但终究还是能够完成的。它表明，五次比较总是足够的，有时必须要比较五次之多，且第二次比较应当在C和D间进行。

这种论证的麻烦在于，要考虑的情况数量会急剧增加。譬如当对象有100个时，需要比较的确切次数是不可能算出来的——几乎可以确定这个数永远无从得知。（我犹记得，当我第一次听一位数学家宣称某个特定量的确切值永远无法算出，我当时有多么震惊。但现在我已经习以为常了，这是常有的事，并不是特殊情况。当时的那个数是拉姆齐数$R(5,5)$，即为了保证n个人中必有五人全部两两认识，同时必有五人全部两两陌生，那么n至少是多大。）作为替代，我们要去寻找上界和下界。对于这个问题来讲，上界c_n的意思是，对n个物体进行排序的某种过程至多进行c_n次比较,下界b_n的意思是:无论你有多聪明,有时候也必须要做上b_n次比较才行。这个例子中,已知的最优上界和下界正相差一个乘数因子:相差常数倍情况下,对n个对象进行排序所需的比较次数为$n\log n$。

要理解这为什么有意思,你不妨自己构思一种排序过程。一种明显的方法是,先找出排序最靠前的对象放到一边,接下来再重复进行。为了找到最好的对象,先比较头两个,再将胜出者与第三个比,再将胜出者与第四个比,依此类推。这种办法下,需要比较$n-1$次才能找到最好的,再找出次好的需要比较$n-2$次,等等等等,于是总的比较次数为$(n-1)+(n-2)+(n-3)+\cdots+1$,计算得约为$n^2/2$。

尽管这种方法很自然,但如果你采用它,最终排列完时就比较了所有对象中的**每**一对,所以它其实是最低效的方法(虽然它的确具有易编程的优点)。当n很大时,$n\log n$是对$n^2/2$很显著的改进,因为$\log n$比$n/2$要小得多。

下述方法被称为"快速排序",它并不能保证一定就会更快,但通常情况它都会快很多。它是如下递归(即利用自身)定义的。首先选择任一对象,比如x,再将其他对象分成两堆,一堆全都优于x,另一堆全都劣于x。这需要比较$n-1$次。接下来你只需分别对两堆进行排序——再次利用快速排序。也就是,对每一堆来讲,选出一个对象,将剩余对象分成两小堆,依此类推。一般来说,除非你运气很差,否则分出的两堆中对象数量是差不多的。于是可以表明,所需比较次数大致为$n\log n$。换句话说,这种方法通常都能以你期待的效率运作,相差常数倍以内。

第八章
常见问题

1. 数学家在30岁以后就不比当年了,这是真的吗?

这种传说影响颇为广泛,正由于人们误解了数学能力的本质,才使得它很有吸引力。人们总喜欢把数学家看作极具天资的人,并认为天资这种东西有些人生来就有,其他人则绝难获得。

其实,年龄与数学成果间的关系对不同人来说差别很大。的确有一部分数学家在20来岁的时候做出了他们最杰出的工作,但绝大多数人都认为,他们的知识水平和专业素质终其一生都在稳健地提高,在许多年里,这种专业水平的增长都能够弥补"原生"脑力的任何衰退(如果确实有"原生"脑力这回事的话)。确实数学家在年逾40岁之后就少有重要的突破性进展了,但这也很有可能是社会学方面的原因。到了40岁时,如果有人还有能力做出突破性的工作,那么他极有可能早已因之前的工作闻名遐迩,因而有可能也不像未成名的年轻数学家那样具有奋斗精神。不过还是有很多反例的,有很多数学家在退休之后热情不减,还继续在数学领域工作。

一般来讲,人们通常所想象的数学家的形象——可能很聪明,但有点古怪,穿着邋遢,毫无性欲,比较孤僻——的确不是一种讨喜的形象。有一部分数学家在一定程度上的确符合这种形象,但如果你认为不这样就做不好数学,这种想法可就太蠢了。实际上,如果所有其他条件都相同的话,可能你还要比这些怪数学家更胜一筹。一开始学习数学的所有学生中,最后成为专职研究人员的比例极小,更多的人在早期阶段便离开了数学,比如失去兴趣、没有申请到读博机会,或者得到了博士学位但没有获得教职。在我看来(实际上不仅只有我这么想),对最终通过了这层层考验的人来说,那些"怪人"所占的比例比占一开始学习数学的学生的比例要小。

对数学家这样的负面刻画可能杀伤力很大,吓走许多本来可能喜欢并且擅长这一领域的人,但是"天才"这个词则更加恶毒,杀伤力更大。这里有一个现成的对"天才"的大致定义:对于别人必须经过多年实践都未必能够掌握的事情,天才就是那些在少年时期就能够轻易做好这些事的人。天才的成就有着魔法般的特质,就好像他们的大脑并不只是比我们更有效率,而是运转方式完全不同。剑桥大学每年都会有一两个数学系本科生,他们经常在数分钟之内就能解决的问题,大多数人——包括应该能够教他们的人——往往需要花上几个小时以上。遇到这种人的时候,我们只能退避三舍、顶礼膜拜。

然而,这些超乎寻常的人并不总是最成功的数学研究者。如果你想要解决某个问题,而之前尝试过的数学家都以失败告终,那么你需要具备种种素质,在这其中天赋(如我所定义的那样)既不是必要的也不是充分的。我们可以通过一个极端一点的例子来说明这一点。安德鲁·怀尔斯(在刚到40岁的时候)证明了费马大定理(即对任意正整数 x, y, z 及大于2的正整数 n, $x^n + y^n$

不可能等于z^n），解决了世界上最著名的数学难题。毫无疑问他很聪明，但他并不是我所说的天才。

你可能会问，如果没有某种神秘的超常脑力，他还可能完成这一切吗？回答是，尽管他的成就非常卓著，但也没有卓越到无法解释的程度。我并不了解究竟是什么因素促使他成功的，但他肯定需要非凡的勇气、坚定和耐心，对他人完成的艰难工作的广泛了解，在正确时间专攻正确领域的运气，以及杰出的战略性眼光。

上面所说的最后一条素质，从根本上要比惊人的大脑运转速度更加重要。数学中绝大多数影响深远的贡献是由"乌龟"们而不是"兔子"们做出的。随着数学家的成长，他们都会逐渐学会这个行当里的各种把戏，部分来自于其他数学家的工作，部分来自于自己对这个问题长时间的思考。是否能将他们的专长用于解决极其困难的问题，则在很大程度上决定于细致的规划：选取一些可能会结出丰硕成果的问题，知道什么时候应该放弃一条思路（相当困难的判断），能够先勾勒出论证问题的大框架继而再时不时地向里面填充细节。这就需要对数学有相当成熟的把握，这绝不与天赋相矛盾，但也并不总是会伴随着天赋。

2. 为什么女性数学家很少见？

真想回避这个问题，因为回答这个问题很容易冒犯别人。但是，在全世界各地的数学系所中，即便是在今日，女性所占比例仍然很小；这是一个值得注意的现象，也是数学生活中的一个重要事实，我被迫感到不得不说点什么，尽管我所要说的也无非是对此感到不解和遗憾。

值得强调的一点是，数学家中女性较少只不过是一种统计现象：确实有十分优秀的女性数学家，与男性同行一样，她们表现优秀的方式也多种多样，有时也包括拥有天赋。没有任何迹象表明，女性在数学中所能达到的成就会有上限。我们有时会读到，在特定的智力测试中——比如说视觉空间能力，男性表现得更优秀，有人认为这解释了他们主导着数学领域的原因。然而，这样的论据不足以令人信服，因为视觉空间能力能够通过练习来增强，而且尽管它有时对数学家有帮助，却并非不可或缺。

更可信的一种理由是社会方面的因素：当男孩子为数学能力感到骄傲时，可以想象某个女孩子可能会为自己擅长这项不那么女性化的事务而感到窘迫。而且，有数学天赋的女孩子所能够效仿的榜样很少，她们只能靠自我保持、自我强化。一项社会因素可能会在之后发挥更大的作用：比起其他学科来，数学需要一个人更加专注，这虽然不是不可能，但也很难与女性的母亲身份相结合。小说家坎迪亚·麦克威廉曾经说，她的每个孩子都使她少写了两本书，不过在几年未动笔之后，她至少还能够重新写小说。但如果你几年没有做数学，你就失去了数学的习惯，很难再重拾了。

有人认为，女性数学家发展起自己事业的时间往往晚于男性同行，而数学家的职业结构倾向于回报早期成就，这就使得女性处于一种不利的地位。最杰出的一些女性数学家的人生故事支持了这种说法。不过她们发展自己职业生涯较晚的原因，基本上都是上面所说的社会原因，而且也有许多这方面的例外。

不过，这些解释看起来都不够充分。我在此不再深入探讨了。我还能做的就是告诉大家,关于这方面已经出了几本书(参见"延伸阅读")。最后再加上一点评论:这样的情况是在不断进步的。数学家中女性所占的比例近年来在稳步提高,而且随着社会大环境的不断改变,这样的现象一定还会持续下去。

3. 数学与音乐息息相通吗?

尽管有很多数学家完全不了解音乐,也很少有音乐家对数学感兴趣,但一直有一种民间观念认为这两个领域是相关联的。其结果就是,当我们听说某位数学家钢琴弹得很好,或者爱好作曲,或者喜欢听巴赫,没有人会对此感到惊奇。

有很多奇闻轶事在讲,各种艺术形式中,数学家为音乐所吸引的最多。也有一些研究声称已经表明,受过音乐教育的儿童在科学领域中表现得更优秀。我们不难猜出为什么会这样。尽管在所有艺术形式中抽象都很重要,但音乐在其中最具有代表性,可以说是最明显的抽象艺术:听音乐所获得的愉悦感,大部分来自于对不具有内在含义的纯粹形式的直接——即使不是完全自觉的——欣赏。

不幸的是,这些传说中的证据很少得到严格的科学支持。关于这种说法,就连应该提出哪些疑问都不好说。如果我们收集到统计数据显著地说明,在相近的社会背景及教育背景下,数学家与其他人相比,弹钢琴的百分比更高,那我们能够从中了解到什么呢? 我自己猜测,**的确会**得到这样的数据。但如果提出一种可经实验验证的理论来说明这其中的关联,会有趣得多。就统计证据而言,如果能够更加详尽明确,也会更有价值。数学和音乐都是内容很广泛的领域,某些人很有可能只对领域中的某一部分有热情,而对其他部分毫无兴趣。数学和音乐趣味之间是否会有微妙的联系？如果有，那将会比这两个领域间整体的粗略相关性更具信息含量。

4. 为什么有那么多人旗帜鲜明地厌恶数学？

我们不常听到别人说他们从来不喜欢生物学，或者英国文学。毫无疑问，并不是所有人都会对这些学科感到兴奋，但是，那些没有热情的人往往完全理解那些有热情的人。相反，数学，以及其他内容高度数学化的学科，诸如物理，似乎不仅仅使人提不起兴趣，而且能激起反感。究竟是什么原因使他们一旦能够抛弃数学时就立刻抛弃，并且一生都对数学心有余悸？

很可能并不是因为数学很无聊，而是数学课的经历很乏味。这一点更容易理解。因为数学总是持续在自身的基础上构建，所以学习时的步步跟进就显得很重要。比方说，如果你不太擅长两位数的乘法，那你很可能就不会对分配律（第二章中讨论过）有良好的直觉。没有这种直觉，你可能就会在计算打开括号$(x+2)(x+3)$时感到不适应，于是你接下来就不能很好地理解二次方程，因而也无法理解为什么黄金分割比是$\frac{1+\sqrt{5}}{2}$。

类似这样的环环相扣还有很多，但是，学习数学时的步步跟进不仅仅是保持技术熟练度而已。数学中常常会引入重要的新思想，新思想会比旧思想更加复杂，每一个新思想的引入都有可能把我们甩在后面。一个很明显的例子就是用字母表示数，很多人对此糊里糊涂，但对某个层次以上的数学来讲这是基础性的。还有其他类似的例子，比如负数、三角函数、指数、对数以及初步的微积分。没有作好准备来进行必要的概念飞跃的人，一旦遇到这些新思想时，就会对其后建立在新思想基础上的一切数学感到并不牢靠。久而久之，他们就会习惯于对数学老师所说的东西仅仅一知半解，日后再错过几次飞跃，恐怕连一知半解也做不到了。同时他们又看到班上其他同学能够轻而易举地跟上课程。因此就不难理解，为什么对许多人来讲数学课成为了一种煎熬。

情况一定是这样的吗？有没有人天生注定就会在学校里厌恶数学，还是说，有可能找出一种不同的数学教学方法，使得排斥数学的人能够大大减少？我相信，小孩子如果在早期接受到热情的好老师一对一教学，长大之后就会喜欢上数学。当然，这并不能直接成为一种可行的教育政策，不过至少告诉我们，数学的教育方法可能有改进空间。

从我在本书中所强调的思想出发，我可以给出一条建议。在上面，我间接地将技术的熟练度与对较难概念的理解作了一番比较，但实际情况似乎是，凡是擅长其中一个方面的必然两个方面都擅长。况且，如果说理解数学对象，大体上就是要学习数学对象所遵从的规则，而非把握其本质，那么我们完全可以预期：技术的熟练度与数学理解力之间并不像我们想象得那样泾渭分明。

这又会对课堂实践产生什么影响呢？我并不赞成革命性的改进——数学教育已经深受其累，我所赞同的是小幅度的改变，有所侧重的小幅变化将会是有益的。比方说，一个小学生犯了个常见错误，觉得 $x^{a+b}=x^a+x^b$。强调表达式 x 内在含义的老师会指出，x^{a+b} 的含义是 $a+b$ 个 x 相乘，显然与 a 个 x 相乘再**乘以** b 个 x 相乘的结果相等。不幸的是，很多孩子觉得这样的论证过于复杂、难以领会，何况一旦 a 和 b 不是正整数，这样的说法就无效了。

如果使用更抽象的办法，那么这些孩子可能会从中获益。正如我在第二章中所指出的，关于指数我们需要了解的一切，都能从几条很简单的规则中推导出来，其中最重要的一条就是$x^{a+b}=x^a x^b$。如果这条规则得到了强调，那么上面的这种错误可能出现的机会就减少了，一旦出现了也很容易纠正：我们只需要提醒犯错的人没有使用正确的规则就行了。当然，熟悉x^3等于x乘以x乘以x这样的基本事实也很重要，但这样的事实可以当作规则的推论出现，而不是当作规则的论据。

我并不是想说，我们应该向孩子们解释什么是抽象方法，我只是想指，教师们应当对抽象方法的隐含意义有所认识。这些隐含意义中最主要的一个就是，即使并不能确切地了解数学概念的含义，我们也很有可能学会正确地使用它们。这听起来似乎是个坏主意，但是用法总是容易教，而对意义的深层理解——倘若在用途之上**的确有**某种意义的话——常常会自然而然地随之而来。

5. 数学家在工作中使用计算机吗？

简单地说，大多数数学家并不用，或者说，即使用也不会在数学工作中占据基础性的地位。当然，正如所有其他人一样，我们也认为，在文字处理和交流沟通方面计算机是不可或缺的，互联网也在日益发挥着重要的作用。在某些数学领域中，人们必须常规性地做冗长烦琐的基本计算，有一些很不错的符号计算程序能够完成这些工作。

所以，计算机能够成为非常有用的节省时间的机器，某些时候，它们甚至能使数学家发现单靠自己无法发现的结果。不过，计算机能够提供的帮助还是很有限的。如果你所研究的问题——更多情况下是子问题——正好是那一小部分能够通过长时间重复性工作完成的，那当然不成问题。可如果你的研究受阻，需要一个聪明的想法，那么就当前的技术状况而言，计算机什么忙也帮不上。实际上，大多数数学家都会说，他们最重要的工具还是纸和笔。

我的观点属于少数派，我认为这种情况只是暂时的，在未来的一百年左右的时间里，计算机将渐渐能够为数学家做越来越多的工作——没准会从帮我们做一些简单的练习开始，或者防止我们在证明错误的引理上浪费一个星期，以免到时才发现一种著名的构造就给出了反例（我说的正是自己常常经历的），直到最终完全取代我们。大多数的数学家则远比我悲观（或者说这才应该属于乐观？），认为计算机很难擅长做数学。

6. 数学研究何以可能进行？

反过来，我们也可以问，数学研究可能进行，这里面哪一点显得很奇怪呢？我在本书中提到过几个未解决的问题，而数学研究在很大程度上就是在努力解决这些问题以及类似问题。如果读过了第七章，那你会看到，有一种生成问题的好办法是去找一种很难精确分析的数学现象，然后努力对它作一些近似的陈述。第六章结尾处还提出了另一种办法：选一种较难的数学概念，诸如四维流形，然后你通常就会发现，关于这些概念，即便很简单的问题也非常难解答。

如果说数学研究中有什么比较神秘的话，那并不是困难问题的存在——实际上，要创造出奇难无比的问题很容易——而是居然有足够的问题恰好有相当的难度，从而钓住数以千计的数学家。要实现这一点，这些问题必然得有挑战性，但同时也必须让人们看到有可能解决的一线希望。

7. 有没有著名数学问题被业余爱好者解决过？

坦率地讲，没有——这就是对这个问题最简单的回答，也是最不具有误导性的回答。专业数学家能够很快地意识到，他们就著名问题所产生的几乎任何思想，都已经有许多前辈想到过了。一种思想要想成为全新的，就必须具备某种特征能够说明为何前人从来没有考虑过它。可能仅仅是这种想法极具原创性，出人意料，但这种情况十分罕见：总体而言，某种思想的诞生会有充足的理由，而不会是凭空冒出来的。如果你有了这种想法，那凭什么别人就不曾有过呢？一种更加合理的理由是，这个想法和别的某种思想相关，那种思想的知名度并不高，而你已经不畏艰难地去学习并且吸收那种思想。这样至少降低了别人在你之前已经有过同样想法的概率，虽然还是没有降到零。

世界各地的数学系所经常收到一些信件，寄信人声称他们解决了某某著名问题，但实际上无一例外，这些"解答"不仅是错误的，而且错得还很滑稽。有一些解答虽然严格地说并不是犯了错误，看起来却根本就不像是在正确地证明任何东西，简直都不能算是去尝试解答。有些解答至少遵循了数学陈述的常规方式，但其中使用的都是非常初等的论据，如果这些论据是正确的，早在许多世纪以前就会被人发现了。写这些信的人对数学研究的艰难程度没有一点概念，不了解要想做出重大的原创性工作，必须要花上数年的时间来充分发展知识和专长，也不知道数学是一项多么需要集体合作的活动。

上面最后一句话并不是指数学家会在很大的团体中工作——尽管很多研究论文的确都署有两到三个作者的名字。我所指的是，随着数学的发展，人们创造出解答某些问题不可或缺的新技术。其结果就是，每一代数学家都站在前一代数学家的肩膀上，去解答过去一度认为难以触及的问题。如果你试图离开数学界主流去孤立地研究问题，那你就必须自行去创造这些技术，这当然会使你处于一种极其不利的境地。

当然我也不是说业余爱好者就不可能在数学中作出重要的研究。实际上的确有一两个这样的例子。1975年，一位几乎没有接受过任何数学训练的圣地亚哥家庭主妇玛乔丽·赖斯，在读过《科学美国人》杂志中谈到的问题后，发现了三种之前不为人知的用（非正）五边形镶嵌平面的办法。在1952年，一位58岁的德国校长库尔特·黑格纳证明了高斯遗留下来一个多世纪的一个著名猜想。

然而，这些例子与我之前所说的并不矛盾。有一些问题看起来与数学的主干联系并不太紧密，对于这些问题来讲，了解现有的数学技术帮助就不是很大。寻找新的五边形镶嵌方式的问题就属于这样的情况：专业数学家去解决它和一位有天资的爱好者去解决它，两者的水平几乎不相上下。赖斯所取得的成绩就好像是一位天文爱好者发现了一颗新的彗星，解决问题为她带来的名声正是她长久研究所应得的回报。至于黑格纳，尽管他不是一位专业数学家，他肯定也不是在孤立地开展工作。尤其是，他自学了模函数的知识。我无法在此解释模函数是什么——实际上，这部分知识通常被视为比数学系的本科课程还要高深。

有意思的是，黑格纳并没有通过完全常规的方式详细写下他的证明。而且尽管他的论文勉强发表了，在很长的时间里也都被当作是错误的。直到1960年代末，这个问题由艾伦·贝克和哈罗德·斯塔克独立地再次解决。直至此时，黑格纳的工作才被仔细地重新检视，并被发现是正确的。不幸的是，黑格纳已经在 1965 年去世，没能亲眼看到自己恢复名誉的那一天。

8. 数学家们为什么会认为某些定理和证明是美丽的？

在本书前面的部分我讨论过这个问题，所以在这里我会说得很简要。用美学的语言来表述数学这一类明显枯燥的事物，这似乎有些奇怪。但正如我在第三章中（在铺地砖问题的结尾处）所说明的，数学论述能够给人愉悦感，这样的愉悦感与更传统的美学愉悦感有很多共同点。

不过，其中一个不同点是——至少在美学观点看来——数学家比艺术家缺少个人特质。我们可能会极为景仰某位发现了美丽证明的数学家，但这项发现背后的人的故事却会最终被淡忘，结果能够愉悦我们的还是数学本身。

索 引

（条目后的数字为原文页码）

A

abstract method, the 抽象方法 18,
20—26, 29—34, 62, 65, 70—71, 82, 84—85, 105, 133

abstraction 抽象 16

approximation 近似 56—57, 61, 63—64,
66—69, 114—115, 121—122, 124

   of logarithms 对数的 116—117

   of multiplication 乘法的 116

   of square roots 平方根的 116

Archimedes 阿基米德 64, 68

area 面积 64—69, 82

   of a circle 圆的 68—69

   of curvilinear shapes 曲线形状的 65—67

   of polygons 多边形的 64—65

associative law, the 结合律 23, 32—33, 62

atlas 地图册

   concept of 的概念 109—111

   of a torus 环面的 109—110

axioms 公理 39—41, 86

   for addition and multiplication 加法和乘法 23, 27

   for area 面积 65

   for inverses 倒数 26—27

   for logarithms 对数 34

   for powers 指数 33

   of Euclidean geometry 欧氏几何 86—87, 90—91, 93, 97—98, 102

B

beauty in mathematics 数学之美 51, 137—138

C

cancellation law, the 消去律 27

Cantor, Georg 格奥尔格·康托尔 40

Cartesian coordinates 笛卡尔坐标 71—74, 92, 97, 106

chess 国际象棋 19—20

commutative law, the 交换律 23, 62

complex numbers 复数 30—31

computers 计算机 11—114, 122, 134

consistency, importance of 自洽性的重要性 32, 41, 71

curved space 弯曲空间 104—108

D

decimal expansions 十进制展开 30, 57—62

Descartes, René 勒内·笛卡尔 71

134

dimension 维度 *see* high-dimensional geometry 参见高维几何

dislike of mathematics 厌恶数学 131—132

distance 距离 73—75, 98—103

distributive law, the 分配律 23

E

Eddington, Sir Arthur 阿瑟·爱丁顿爵士 107

Einstein, Albert 阿尔伯特·爱因斯坦 107

Euclid 欧几里得 56, 86, 90

Euclidean geometry 欧氏几何 *see* geometry, Euclidean 参见几何,欧氏

Euclid's *Elements* 欧几里得的《几何原本》 86

exponentiation 指数幂 117

F

Fermat's Last Theorem 费马大定理 128

fractional dimension 分数维度 82—85

fractions 分数 *see* rational numbers 参见有理数

Frege, Gottlob 戈特洛布·弗雷格 40

fundamental theorem of algebra, the 代数基本定理 31

fundamental theorem of arithmetic, the 算术基本定理 52

G

Galileo 伽利略 3

Gauss, Carl Friedrich 卡尔·弗雷德里希·高斯 31, 90, 107, 137

genius 天才 127—128

geometry 几何
    Euclidean 欧氏 86—91
    hyperbolic 双曲 98—104, 107—108
    spherical 球面 93—98

Goldbach's conjecture 哥德巴赫猜想 121

golden ratio, the 黄金分割比 42—45

graphs 图 15—16, 19

great circles 大圆 93—95, 97

H

Heegner, Kurt 库尔特·黑格纳 137

high-dimensional geometry 高维几何 70—80
    cubes and spheres in 中的立方体与球面 75—79
    reason for studying 研究的理由 80—81

Hilbert, David 大卫·希尔伯特 ix
Hilbert spaces 希尔伯特空间 ix-x
hyperbolic geometry 双曲几何 see
 geometry,hyperbolic 参见 几何,
 双曲

I

induction, principle of 归纳法原理
 39, 41
infinity 无穷 32, 56—62, 64
instantaneous speed 瞬时速度 62—64
irrational numbers 无理数 29, 36

J

Jordan curve theorem, the 若尔当曲
 线定理 54

K

Kant, Immanuel 伊曼努尔·康德 90
Koch snowflake, the 科赫雪花 83—85

L

logarithms 对数 34, 116—117, 119, 124—
 125

M

manifolds 流形 108—111
mathematical models 数学模型 3—5,
 16, 28, 31, 71, 80
 of brains 大脑的 14
 of computers 计算机的 11—14
 of dice 掷骰子的 5—6
 of gases 气体的 8—11
 of high-dimensional space 高维空
  间的 71—75
 of population growth 人口增长的
  6—7
 of timetabling 时间表的 14—15
Mercator's projection 墨卡托投影
 100—101
music and mathematics 音乐与数学
 51, 130—131

N

natural numbers 自然数 20—24
negative numbers 负数 26—28
number systems 数系 21—22, 26, 28—29

P

parallel postulate, the 平行公设 87—

88, 102—104

    attempted proofs of 的尝试证明 91—98,104

Poincaré, Henri 亨利·庞加莱 98

Poincaré conjecture, the 庞加莱猜想 111

prime number 素数 118—122

    theorem, the 定理 119

probability 概率 5—6, 8, 10,118—121

proof 证明 36, 39—42, 48, 50—51

    by contradiction 反证法 36

    of Pythagoras' theorem 毕达哥拉斯定理的 48—49

    that the golden ratio is irrational 黄金分割比为无理数的 41—45

    that the square root of two is irrational 根号二为无理数的 36—39

Pythagoras 毕达哥拉斯 29

    theorem of 定理 48—49, 73—74

Q

qualia 感受性 31

Quicksort 快速排序 125

R

raising numbers to powers 指数运算 33—34, 133

rational numbers 有理数 26—28,36—37

real numbers 实数 29

reductio ad absurdum 归谬法 *see* proof by contradiction 参见证明—反证法

reinterpretation 重新解释 97—98, 103

Rice, Marjorie 玛乔丽·赖斯 136—137

Riemann hypothesis, the 黎曼假设 121—122

Russell, Bertrand 伯特兰·罗素 40, 140

S

Saussure, Ferdinand de 费尔迪南·德·索绪尔 18, 139

simplifying assumptions 简化假设 2—3, 6—9

sorting algorithms 排序算法 122—125

spherical geometry 球面几何 *see* geometry, spherical 参见几何，球面

square root of minus one, the 负一的平方根 *see* complex numbers 参见复数

square root of two, the 根号二 29—30, 36—38, 56—62

T

teaching of mathematics 数学教育 132—133

tiling problem 铺地砖问题 49—51

trefoil knot, the 三叶结 52—53

V

Vinogradov, Ivan 伊万·维诺格拉多夫 120—121

visualization 图像化 78—80

volume 体积 82—83

W

Whitehead, Alfred North 阿尔弗雷德·诺斯·怀特海 40,140

Wiles, Andrew 安德鲁·怀尔斯 127—128

Wittgenstein, Ludwig 路德维希·维特根斯坦 18, 28,139

women in mathematics 数学领域的女性 128—130, 140

Z

zero 零 24—25

division by 作除数 26, 32

Timothy Gowers

MATHEMATICS

A Very Short Introduction

Contents

Preface ix

List of diagrams xiii

1 Models 1
2 Numbers and abstraction 17
3 Proofs 35
4 Limits and infinity 56
5 Dimension 70
6 Geometry 86
7 Estimates and approximations 112
8 Some frequently asked questions 126

Further reading 139

Preface

Early in the 20th century, the great mathematician David Hilbert noticed that a number of important mathematical arguments were structurally similar. In fact, he realized that at an appropriate level of generality they could be regarded as the same. This observation, and others like it, gave rise to a new branch of mathematics, and one of its central concepts was named after Hilbert. The notion of a Hilbert space sheds light on so much of modern mathematics, from number theory to quantum mechanics, that if you do not know at least the rudiments of Hilbert space theory then you cannot claim to be a well-educated mathematician.

What, then, is a Hilbert space? In a typical university mathematics course it is defined as a complete inner-product space. Students attending such a course are expected to know, from previous courses, that an inner-product space is a vector space equipped with an inner product, and that a space is complete if every Cauchy sequence in it converges. Of course, for those definitions to make sense, the students also need to know the definitions of vector space, inner product, Cauchy sequence and convergence. To give just one of them (not the longest): a Cauchy sequence is a sequence $x_1, x_2, x_3, \ldots$ such that for every positive number ϵ there exists an integer N such that for any two integers p and q greater than N the distance from x_p to x_q is at most ϵ.

In short, to have any hope of understanding what a Hilbert space is, you must learn and digest a whole hierarchy of lower-level concepts first. Not surprisingly, this takes time and effort. Since the same is true of many of the most important mathematical ideas, there is a severe limit to what can be achieved by any book that attempts to offer an accessible introduction to mathematics, especially if it is to be very short.

Instead of trying to find a clever way round this difficulty, I have focused on a different barrier to mathematical communication. This one, which is more philosophical than technical, separates those who are happy with notions such as infinity, the square root of minus one, the twenty-sixth dimension, and curved space from those who find them disturbingly paradoxical. It is possible to become comfortable with these ideas without immersing oneself in technicalities, and I shall try to show how.

If this book can be said to have a message, it is that one should learn to think abstractly, because by doing so many philosophical difficulties simply disappear. I explain in detail what I mean by the abstract method in Chapter 2. Chapter 1 concerns a more familiar, and related, kind of abstraction: the process of distilling the essential features from a real-world problem, and thereby turning it into a mathematical one. These two chapters, and Chapter 3, in which I discuss what is meant by a rigorous proof, are about mathematics in general.

Thereafter, I discuss more specific topics. The last chapter is more about mathematicians than about mathematics and is therefore somewhat different in character from the others. I recommend reading Chapter 2 before the later ones, but apart from that the book is arranged as unhierarchically as possible: I shall not assume, towards the end of the book, that the reader has understood and remembered everything that comes earlier.

Very little prior knowledge is needed to read this book – a British GCSE course or its equivalent should be enough – but I do presuppose some interest on the part of the reader rather than trying to drum it up myself.

For this reason I have done without anecdotes, cartoons, exclamation marks, jokey chapter titles, or pictures of the Mandelbrot set. I have also avoided topics such as chaos theory and Gödel's theorem, which have a hold on the public imagination out of proportion to their impact on current mathematical research, and which are in any case well treated in many other books. Instead, I have taken more mundane topics and discussed them in detail in order to show how they can be understood in a more sophisticated way. In other words, I have aimed for depth rather than breadth, and have tried to convey the appeal of mainstream mathematics by letting it speak for itself.

I would like to thank the Clay Mathematics Institute and Princeton University for their support and hospitality during part of the writing of the book. I am very grateful to Gilbert Adair, Rebecca Gowers, Emily Gowers, Patrick Gowers, Joshua Katz, and Edmund Thomas for reading earlier drafts. Though they are too intelligent and well informed to count as general readers, it is reassuring to know that what I have written is comprehensible to at least some non-mathematicians. Their comments have resulted in many improvements. To Emily I dedicate this book, in the hope that it will give her a small idea of what it is I do all day.

List of diagrams

1. A ball in flight I — 4
2. A ball in flight II — 4
   © PhotoDisc
3. A two-dimensional model of a gas — 9
4. A primitive computer program — 12
5. A graph with 10 vertices and 15 edges — 15
6. White starts, and has a winning strategy — 19
7. The concept of fiveness — 20
8. Ways of representing 7, 12, and 47 (twice) — 21
9. Dividing a circle into regions — 35
10. The existence of the golden ratio — 42
11. Cutting squares off rectangles — 44
12. Regions of a circle — 47
13. A short proof of Pythagoras' theorem — 49
14. Tiling a square grid with the corners removed — 50
15. A trefoil knot — 53
16. Four kinds of curve — 54
17. Is the black spot inside the curve or outside it? — 55
18. Approximating the area of a curved shape — 66
19. Archimedes' method for showing that a circle has area πr^2 — 68

20	Approximating a circle by a polygon	69	29	The uniqueness of parallel lines	92

20 Approximating a circle by a polygon 69

21 Three points in the Cartesian plane 72

22 Calculating distances using Pythagoras' theorem 73

23 A unit square and a unit cube 79

24 Dividing a square into $9 = 3^2$ smaller squares, and a cube into $27 = 3^3$ smaller cubes 83

25 Building up the Koch snowflake 84

26 Euclid's fourth axiom and two versions of the fifth 88

27 A consequence of Euclid's fifth axiom 89

28 A proof that the angles of a triangle add up to 180 degrees 89

29 The uniqueness of parallel lines 92

30 A great circle 95

31 The parallel postulate is false for spherical geometry 96

32 The phrase 'in the same direction as' does not make sense on the surface of a sphere 96

33 A tessellation of the hyperbolic plane by regular pentagons 99

34 A typical hyperbolic line 101

35 A typical hyperbolic circle, and its centre 103

36 The parallel postulate is false in the hyperbolic plane 104

37 A hyperbolic triangle 108

38 An atlas of a torus 110

The publisher and the author apologize for any errors or omissions in the above list. If contacted they will be pleased to rectify these at the earliest opportunity.

Chapter 1
Models

How to throw a stone

Suppose that you are standing on level ground on a calm day, and have in your hand a stone which you would like to throw as far as possible. Given how hard you can throw, the most important decision you must make is the angle at which the stone leaves your hand. If this angle is too flat, then although the stone will have a large horizontal speed it will land quite soon and will therefore not have a chance to travel very far. If on the other hand you throw the stone too high, then it will stay in the air for a long time but without covering much ground in the process. Clearly some sort of compromise is needed.

The best compromise, which can be worked out using a combination of Newtonian physics and some elementary calculus, turns out to be as neat as one could hope for under the circumstances: the direction of the stone as it leaves your hand should be upwards at an angle of 45 degrees to the horizontal. The same calculations show that the stone will trace out a parabolic curve as it flies through the air, and they tell you how fast it will be travelling at any given moment after it leaves your hand.

It seems, therefore, that a combination of science and mathematics enables one to predict the entire behaviour of the stone from the moment it is launched until the moment it lands. However, it does so only if one is prepared to make a number of simplifying assumptions, the main one being that the only force acting on the stone is the earth's gravity and that this force has the same magnitude and direction everywhere. That is not true, though, because it fails to take into account air resistance, the rotation of the earth, a small gravitational influence from the moon, the fact that the earth's gravitational field is weaker the higher you are, and the gradually changing direction of 'vertically downwards' as you move from one part of the earth's surface to another. Even if you accept the calculations, the recommendation of 45 degrees is based on another implicit assumption, namely that the speed of the stone as it leaves your hand does not depend on its direction. Again, this is untrue: one can throw a stone harder when the angle is flatter.

In the light of these objections, some of which are clearly more serious than others, what attitude should one take to the calculations and the predictions that follow from them? One approach would be to take as many of the objections into account as possible. However, a much more sensible policy is the exact opposite: decide what level of accuracy you need, and then try to achieve it as simply as possible. If you know from experience that a simplifying assumption will have only a small effect on the answer, then you should make that assumption.

For example, the effect of air resistance on the stone will be fairly small because the stone is small, hard, and reasonably dense. There is not much point in complicating the calculations by taking air resistance into account when there is likely to be a significant error in the angle at which one ends up throwing the stone anyway. If you want to take it into account, then for most purposes the following rule of thumb is good enough: the greater the air resistance, the flatter you should make your angle to compensate for it.

What is a mathematical model?

When one examines the solution to a physical problem, it is often, though not always, possible to draw a clear distinction between the contributions made by science and those made by mathematics. Scientists devise a theory, based partly on the results of observations and experiments, and partly on more general considerations such as simplicity and explanatory power. Mathematicians, or scientists doing mathematics, then investigate the purely logical consequences of the theory. Sometimes these are the results of routine calculations that predict exactly the sorts of phenomena the theory was designed to explain, but occasionally the predictions of a theory can be quite unexpected. If these are later confirmed by experiment, then one has impressive evidence in favour of the theory.

The notion of confirming a scientific prediction is, however, somewhat problematic, because of the need for simplifications of the kind I have been discussing. To take another example, Newton's laws of motion and gravity imply that if you drop two objects from the same height then they will hit the ground (if it is level) at the same time. This phenomenon, first pointed out by Galileo, is somewhat counter-intuitive. In fact, it is worse than counter-intuitive: if you try it for yourself, with, say, a golf ball and a table-tennis ball, you will find that the golf ball lands first. So in what sense was Galileo correct?

It is, of course, because of air resistance that we do not regard this little experiment as a refutation of Galileo's theory: experience shows that the theory works well when air resistance is small. If you find it too convenient to let air resistance come to the rescue every time the predictions of Newtonian mechanics are mistaken, then your faith in science, and your admiration for Galileo, will be restored if you get the chance to watch a feather fall in a vacuum – it really does just drop as a stone would.

Nevertheless, because scientific observations are never completely direct and conclusive, we need a better way to describe the relationship between science and mathematics. Mathematicians do not apply scientific theories directly to the world but rather to *models*. A model in this sense can be thought of as an imaginary, simplified version of the part of the world being studied, one in which exact calculations are possible. In the case of the stone, the relationship between the world and the model is something like the relationship between Figures 1 and 2.

1. A ball in flight I

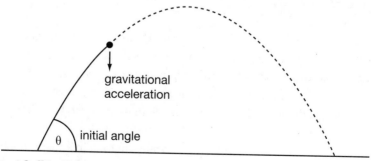

2. A ball in flight II

There are many ways of modelling a given physical situation, and we must use a mixture of experience and further theoretical considerations to decide what a given model is likely to teach us about the world itself. When choosing a model, one priority is to make its behaviour correspond closely to the actual, observed behaviour of the world. However, other factors, such as simplicity and mathematical elegance, can often be more important. Indeed, there are very useful models with almost no resemblance to the world at all, as some of my examples will illustrate.

Rolling a pair of dice

If I roll a pair of dice and want to know how they will behave, then experience tells me that there are certain questions it is unrealistic to ask. For example, nobody could be expected to tell me the outcome of a given roll in advance, even if they had expensive technology at their disposal and the dice were to be rolled by a machine. By contrast, questions of a probabilistic nature, such as, 'How likely is it that the numbers on the dice will add up to seven?' can often be answered, and the answers may be useful if, for example, I am playing backgammon for money. For the second sort of question, one can model the situation very simply by representing a roll of the dice as a random choice of one of the following thirty-six pairs of numbers.

$$
\begin{array}{cccccc}
(1,1) & (1,2) & (1,3) & (1,4) & (1,5) & (1,6) \\
(2,1) & (2,2) & (2,3) & (2,4) & (2,5) & (2,6) \\
(3,1) & (3,2) & (3,3) & (3,4) & (3,5) & (3,6) \\
(4,1) & (4,2) & (4,3) & (4,4) & (4,5) & (4,6) \\
(5,1) & (5,2) & (5,3) & (5,4) & (5,5) & (5,6) \\
(6,1) & (6,2) & (6,3) & (6,4) & (6,5) & (6,6)
\end{array}
$$

The first number in each pair represents the number showing on the first die, and the second the number on the second. Since exactly six of the pairs consist of two numbers that add up to seven, the chances of rolling a seven are six in thirty-six, or one in six.

One might object to this model on the grounds that the dice, when rolled, are obeying Newton's laws, at least to a very high degree of precision, so the way they land is anything but random: indeed, it could in principle be calculated. However, the phrase 'in principle' is being overworked here, since the calculations would be extraordinarily complicated, and would need to be based on more precise information about the shape, composition, initial velocities, and rotations of the dice than could ever be measured in practice. Because of this, there is no advantage whatsoever in using some more complicated deterministic model.

Predicting population growth

The 'softer' sciences, such as biology and economics, are full of mathematical models that are vastly simpler than the phenomena they represent, or even deliberately inaccurate in certain ways, but nevertheless useful and illuminating. To take a biological example of great economic importance, let us imagine that we wish to predict the population of a country in 20 years' time. One very simple model we might use represents the entire country as a pair of numbers $(t, P(t))$. Here, t represents the time and $P(t)$ stands for the size of the population at time t. In addition, we have two numbers, b and d, to represent birth and death rates. These are defined to be the number of births and deaths per year, as a proportion of the population.

Suppose we know that the population at the beginning of the year 2002 is P. According to the model just defined, the number of births and deaths during the year will be bP and dP respectively, so the population at the beginning of 2003 will be $P + bP - dP = (1 + b - d)P$. This argument works for any year, so we have the formula $P(n + 1) = (1 + b - d)P(n)$, meaning that the population at the beginning of year $n + 1$ is $(1 + b - d)$ times the population at the beginning of year n. In other words, each year the population multiplies by $(1 + b - d)$. It follows that in 20 years it multiplies by $(1 + b - d)^{20}$, which gives an answer to our original question.

Even this basic model is good enough to persuade us that if the birth rate is significantly higher than the death rate, then the population will grow extremely rapidly. However, it is also unrealistic in ways that can make its predictions very inaccurate. For example, the assumption that birth and death rates will remain the same for 20 years is not very plausible, since in the past they have often been affected by social changes and political events such as improvements in medicine, new diseases, increases in the average age at which women start to have children, tax incentives, and occasional large-scale wars. Another reason to expect birth and death rates to vary over time is that the ages of people in the country may be distributed rather unevenly. For example, if there has been a baby boom 15 years earlier, then there is some reason to expect the birth rate to rise in 10 to 15 years' time.

It is therefore tempting to complicate the model by introducing other factors. One could have birth and death rates $b(t)$ and $d(t)$ that varied over time. Instead of a single number $P(t)$ representing the size of the population, one might also like to know how many people there are in various age groups. It would also be helpful to know as much as possible about social attitudes and behaviour in these age groups in order to predict what future birth and death rates are likely to be. Obtaining this sort of statistical information is expensive and difficult, but the information obtained can greatly improve the accuracy of one's predictions. For this reason, no single model stands out as better than all others. As for social and political changes, it is impossible to say with any certainty what they will be. Therefore the most that one can reasonably ask of any model is predictions of a conditional kind: that is, ones that tell us what the effects of social and political changes will be if they happen.

The behaviour of gases

According to the kinetic theory of gases, introduced by Daniel Bernoulli in 1738 and developed by Maxwell, Boltzmann, and others in the second half of the 19th century, a gas is made up of moving molecules, and many of its properties, such as temperature and pressure, are statistical properties of those molecules. Temperature, for example, corresponds to their average speed.

With this idea in mind, let us try to devise a model of a gas contained in a cubical box. The box should of course be represented by a cube (that is, a mathematical rather than physical one), and since the molecules are very small it is natural to represent them by points in the cube. These points are supposed to move, so we must decide on the rules that govern how they move. At this point we have to make some choices.

If there were just one molecule in the box, then there would be an obvious rule: it travels at constant speed, and bounces off the walls of the box when it hits them. The simplest conceivable way to generalize this model is then to take N molecules, where N is some large number, and assume that they all behave this way, with absolutely no interaction between them. In order to get the N-molecule model started, we have to choose initial positions and velocities for the molecules, or rather the points representing them. A good way of doing this is to make the choice randomly, since we would expect that at any given time the molecules in a real gas would be spread out and moving in many directions.

It is not hard to say what is meant by a random point in the cube, or a random direction, but it is less clear how to choose a speed randomly, since speed can take any value from 0 to infinity. To avoid this difficulty, let us make the physically implausible assumption that all the molecules are moving at the same speed, and that it is only the initial positions and directions that are chosen randomly. A

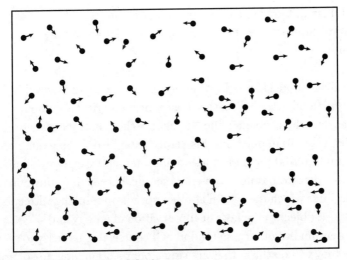

3. A two-dimensional model of a gas

two-dimensional version of the resulting model is illustrated in Figure 3.

The assumption that our N molecules move entirely independently of one another is quite definitely an oversimplification. For example, it means that there is no hope of using this model to understand why a gas becomes a liquid at sufficiently low temperatures: if you slow down the points in the model you get the same model, but running more slowly. Nevertheless, it does explain much of the behaviour of real gases. For example, imagine what would happen if we were gradually to shrink the box. The molecules would continue to move at the same speed, but now, because the box was smaller, they would hit the walls more often and there would be less wall to hit. For these two reasons, the number of collisions per second in any given area of wall would be greater. These collisions account for the pressure that a gas exerts, so we can conclude that if you squeeze a gas into a smaller volume, then its pressure is likely to increase – as is confirmed by observation. A similar argument explains why, if you increase the temperature of a gas without increasing its volume, its pressure also increases.

And it is not too hard to work out what the numerical relationships between pressure, temperature, and volume should be.

The above model is roughly that of Bernoulli. One of Maxwell's achievements was to discover an elegant theoretical argument that solves the problem of how to choose the initial speeds more realistically. To understand this, let us begin by dropping our assumption that the molecules do not interact. Instead, we shall assume that from time to time they collide, like a pair of tiny billiard balls, after which they go off at other speeds and in other directions that are subject to the laws of conservation of energy and momentum but otherwise random. Of course, it is not easy to see how they will do this if they are single points occupying no volume, but this part of the argument is needed only as an informal justification for some sort of randomness in the speeds and directions of the molecules. Maxwell's two very plausible assumptions about the nature of this randomness were that it should not change over time and that it should not distinguish between one direction and another. Roughly speaking, the second of these assumptions means that if d_1 and d_2 are two directions and s is a certain speed, then the chances that a particle is travelling at speed s in direction d_1 are the same as the chances that it is travelling at speed s in direction d_2. Surprisingly, these two assumptions are enough to determine exactly how the velocities should be distributed. That is, they tell us that if we want to choose the velocities randomly, then there is only one natural way to do it. (They should be assigned according to the normal distribution. This is the distribution that produces the famous 'bell curve', which occurs in a large number of different contexts, both mathematical and experimental.)

Once we have chosen the velocities, we can again forget all about interactions between the molecules. As a result, this slightly improved model shares many of the defects of the first one. In order to remedy them, there is no choice but to model the interactions somehow. It turns out that even very simple models of systems of interacting particles behave in a fascinating way and give rise to extremely difficult, indeed mostly unsolved, mathematical problems.

Modelling brains and computers

A computer can also be thought of as a collection of many simple parts that interact with one another, and largely for this reason theoretical computer science is also full of important unsolved problems. A good example of the sort of question one might like to answer is the following. Suppose that somebody chooses two prime numbers p and q, multiplies them together and tells you the answer pq. You can then work out what p and q are by taking every prime number in turn and seeing whether it goes exactly into pq. For example, if you are presented with the number 91, you can quickly establish that it is not a multiple of 2, 3, or 5, and then that it equals 7×13.

If, however, p and q are very large – with 200 digits each, say – then this process of trial and error takes an unimaginably long time, even with the help of a powerful computer. (If you want to get a feel for the difficulty, try finding two prime numbers that multiply to give 6901 and another two that give 280123.) On the other hand, it is not inconceivable that there is a much cleverer way to approach the problem, one that might be used as the basis for a computer program that does not take too long to run. If such a method could be found, it would allow one to break the codes on which most modern security systems are based, on the Internet and elsewhere, since the difficulty of deciphering these codes depends on the difficulty of factorizing large numbers. It would therefore be reassuring if there were some way of showing that a quick, efficient procedure for calculating p and q from their product pq does not exist. Unfortunately, while computers continually surprise us with what they can be used for, almost nothing is known about what they cannot do.

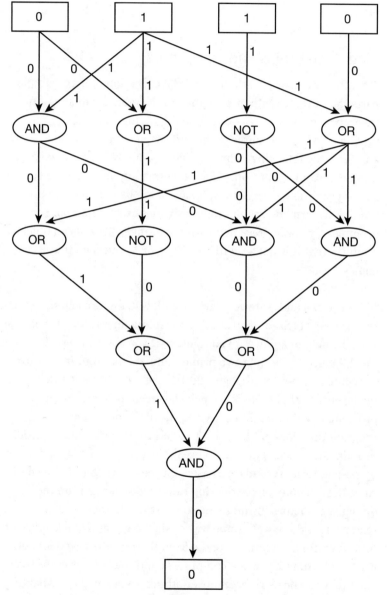

4. A primitive computer program

Before one can begin to think about this problem one must find some way of representing a computer mathematically, and as simply as possible. Figure 4 shows one of the best ways of doing this. It consists of layers of nodes that are linked to one another by lines that are called edges. Into the top layer goes the 'input', which is a sequence of 0s and 1s, and out of the bottom layer comes the 'output', which is another sequence of 0s and 1s. The nodes are of three kinds, called AND, OR, and NOT gates. Each of these gates receives some 0s and 1s from the edges that enter it from above. Depending on what it receives, it then sends out 0s or 1s itself, according to the following simple rules: if an AND gate receives nothing but 1s then it sends out 1s, and otherwise it sends out 0s; if an OR gate receives nothing but 0s then it sends out 0s, and otherwise it sends out 1s; only one edge is allowed to enter a NOT gate from above, and it sends out 1s if it receives a 0 and 0s if it receives a 1.

An array of gates linked by edges is called a *circuit*, and what I have described is the circuit model of computation. The reason 'computation' is an appropriate word is that a circuit can be thought of as taking one sequence of 0s and 1s and transforming it into another, according to some predetermined rules which may, if the circuit is large, be very complicated. This is also what computers do, although they translate these sequences out of and into formats that we can understand, such as high-level programming languages, windows, icons, and so on. There turns out to be a fairly simple way (from a theoretical point of view – it would be a nightmare to do in practice) of converting any computer program into a circuit that transforms 01-sequences according to exactly the same rules. Moreover, important characteristics of computer programs have their counterparts in the resulting circuits.

In particular, the number of nodes in the circuit corresponds to the length of time the computer program takes to run. Therefore, if one can show that a certain way of transforming 01-sequences needs a very large circuit, then one has also shown that it needs a computer program that runs for a very long time. The advantage of using the circuit model over analysing computers directly is that, from the mathematical point of view, circuits are simpler, more natural, and easier to think about.

A small modification to the circuit model leads to a useful model of the brain. Now, instead of 0s and 1s, one has signals of varying strengths that can be represented as numbers between 0 and 1. The gates, which correspond to neurons, or brain cells, are also different, but they still behave in a very simple way. Each one receives some signals from other gates. If the total strength of these signals – that is, the sum of all the corresponding numbers – is large enough, then the gate sends out its own signals of certain strengths. Otherwise, it does not. This corresponds to the decision of a neuron whether or not to 'fire'.

It may seem hard to believe that this model could capture the full complexity of the brain. However, that is partly because I have said nothing about how many gates there should be or how they should be arranged. A typical human brain contains about 100 billion neurons arranged in a very complicated way, and in the present state of knowledge about the brain it is not possible to say all that much more, at least about the fine detail. Nevertheless, the model provides a useful theoretical framework for thinking about how the brain might work, and it has allowed people to simulate certain sorts of brain-like behaviour.

Colouring maps and drawing up timetables

Suppose that you are designing a map that is divided into regions, and you wish to choose colours for the regions. You would like to use as few colours as possible, but do not wish to give two adjacent regions the same colour. Now suppose that you are drawing up the timetable for a university course that is divided into modules. The number of possible times for lectures is limited, so some modules will have to clash with others. You have a list of which students are taking which modules, and would like to choose the times in such a way that two modules clash only when there is nobody taking both.

These two problems appear to be quite different, but an appropriate choice of model shows that from the mathematical point of view they are the same. In both cases there are some objects (countries, modules) to which something must be assigned (colours, times). Some pairs of objects are incompatible (neighbouring countries, modules that must not clash) in the sense that they are not allowed to receive the same assignment. In neither problem do we really care what the objects are or what is being assigned to them, so we may as well just represent them as points. To show which pairs of points are incompatible we can link them with lines. A collection of points, some of which are joined by lines, is a mathematical structure known as a graph. Figure 5 gives a simple example. It is customary to call the points in a graph vertices, and the lines edges.

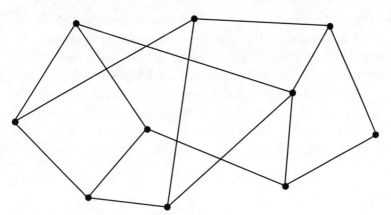

5. **A graph with 10 vertices and 15 edges**

Once we have represented the problems in this way, our task in both cases is to divide the vertices into a small number of groups in such a way that no group contains two vertices linked by an edge. (The graph in Figure 5 can be divided into three such groups, but not into two.) This illustrates another very good reason for making models as simple as possible: if you are lucky, the same model can be used to study many different phenomena at once.

Various meanings of the word 'abstract'

When devising a model, one tries to ignore as much as possible about the phenomenon under consideration, abstracting from it only those features that are essential to understanding its behaviour. In the examples I have discussed, stones were reduced to single points, the entire population of a country to one number, the brain to a network of gates obeying very simple mathematical rules, and the interactions between molecules to nothing at all. The resulting mathematical structures were abstract representations of the concrete situations being modelled.

These are two senses in which mathematics is an abstract subject: it abstracts the important features from a problem and it deals with objects that are not concrete and tangible. The next chapter will discuss a third, deeper sense of abstraction in mathematics, of which the example of the previous section has already given us some idea. A graph is a very flexible model with many uses. However, when one studies graphs, there is no need to bear these uses in mind: it does not matter whether the points represent regions, lectures, or something quite different again. A graph theorist can leave behind the real world entirely and enter the realm of pure abstraction.

Chapter 2
Numbers and abstraction

The abstract method

A few years ago, a review in the *Times Literary Supplement* opened with the following paragraph:

> Given that $0 \times 0 = 0$ and $1 \times 1 = 1$, it follows that there are numbers that are their own squares. But then it follows in turn that there are numbers. In a single step of artless simplicity, we seem to have advanced from a piece of elementary arithmetic to a startling and highly controversial philosophical conclusion: that numbers exist. You would have thought that it should have been more difficult.
>
> A. W. Moore reviewing *Realistic Rationalism*,
> by Jerrold J. Katz, in the *T.L.S.*, 11th September 1998.

This argument can be criticized in many ways, and it is unlikely that anybody takes it seriously, including the reviewer. However, there certainly are philosophers who take seriously the question of whether numbers exist, and this distinguishes them from mathematicians, who either find it obvious that numbers exist or do not understand what is being asked. The main purpose of this chapter is to explain why it is that mathematicians can, and even should, happily ignore this seemingly fundamental question.

The absurdity of the 'artlessly simple' argument for the existence of numbers becomes very clear if one looks at a parallel argument about the game of chess. Given that the black king, in chess, is sometimes allowed to move diagonally by one square, it follows that there are chess pieces that are sometimes allowed to move diagonally by one square. But then it follows in turn that there are chess pieces. Of course, I do not mean by this the mundane statement that people sometimes build chess sets – after all, it is possible to play the game without them – but the far more 'startling' philosophical conclusion that chess pieces exist independently of their physical manifestations.

What is the black king in chess? This is a strange question, and the most satisfactory way to deal with it seems to be to sidestep it slightly. What more can one do than point to a chessboard and explain the rules of the game, perhaps paying particular attention to the black king as one does so? What matters about the black king is not its existence, or its intrinsic nature, but the role that it plays in the game.

The abstract method in mathematics, as it is sometimes called, is what results when one takes a similar attitude to mathematical objects. This attitude can be encapsulated in the following slogan: a mathematical object *is* what it *does*. Similar slogans have appeared many times in the philosophy of language, and can be highly controversial. Two examples are 'In language there are only differences' and 'The meaning of a word is its use in the language', due to Saussure and Wittgenstein respectively (see Further reading), and one could add the rallying cry of the logical positivists: 'The meaning of a statement is its method of verification.' If you find mine unpalatable for philosophical reasons, then, rather than regarding it as a dogmatic pronouncement, think of it as an attitude which one can sometimes choose to adopt. In fact, as I hope to demonstrate, it is essential to be able to adopt it if one wants a proper understanding of higher mathematics.

Chess without the pieces

It is amusing to see, though my argument does not depend on it, that chess, or any other similar game, can be modelled by a graph. (Graphs were defined at the end of the previous chapter.) The vertices of the graph represent possible positions in the game. Two vertices P and Q are linked by an edge if the person whose turn it is to play in position P has a legal move that results in position Q. Since it may not be possible to get back from Q to P again, the edges need arrows on them to indicate their direction. Certain vertices are considered wins for white, and others wins for black. The game begins at one particular vertex, corresponding to the starting position of the game. Then the players take turns to move forwards along edges. The first player is trying to reach one of white's winning vertices, and the second one of black's. A far simpler game of this kind is illustrated in Figure 6. (It is not hard to see that for this game white has a winning strategy.)

This graph model of chess, though wildly impractical because of the vast number of possible chess positions, is perfect, in the sense that the resulting game is exactly equivalent to chess. And yet, when I defined it I made no mention of chess pieces at all. From this perspective, it seems quite extraordinary to ask whether the black

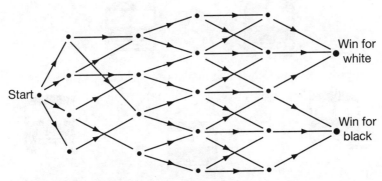

6. White starts, and has a winning strategy

king exists: the chessboard and pieces are nothing more than convenient organizing principles that help us think about the bewildering array of vertices and edges in a huge graph. If we say something like, 'The black king is in check', then this is just an abbreviation of a sentence that specifies an extremely long list of vertices and tells us that the players have reached one of them.

The natural numbers

'Natural' is the name given by mathematicians to the familiar numbers 1,2,3,4, ... They are among the most basic of mathematical objects, but they do not seem to encourage us to think abstractly. What, after all, could a number like 5 be said to *do*? It doesn't move around like a chess piece. Instead, it seems to have an intrinsic nature, a sort of pure fiveness that we immediately grasp when we look at a picture such as Figure 7.

However, when we consider larger numbers, there is rather less of this purity. Figure 8 gives us representations of the numbers 7, 12, and 47. Perhaps some people instantly grasp the sevenness of the first picture, but in most people's minds there will be a fleeting

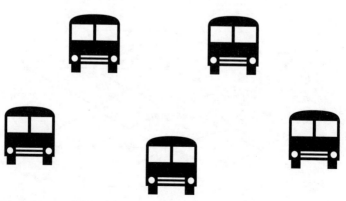

7. The concept of fiveness

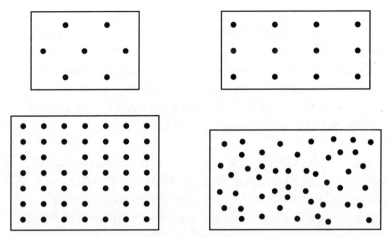

8. Ways of representing 7, 12, and 47 (twice)

thought such as, 'The outer dots form a hexagon, so together with the central one we get $6 + 1 = 7$.' Likewise, 12 will probably be thought of as 3×4, or 2×6. As for 47, there is nothing particularly distinctive about a group of that number of objects, as opposed to, say, 46. If they are arranged in a pattern, such as a 7×7 grid with two points missing, then we can use our knowledge that $7 \times 7 - 2 = 49 - 2 = 47$ to tell quickly how many there are. If not, then we have little choice but to count them, this time thinking of 47 as the number that comes after 46, which itself is the number that comes after 45, and so on.

In other words, numbers do not have to be very large before we stop thinking of them as isolated objects and start to understand them through their properties, through how they relate to other numbers, through their role in a *number system*. This is what I mean by what a number 'does'.

As is already becoming clear, the concept of a number is intimately connected with the arithmetical operations of addition and multiplication: for example, without some idea of arithmetic one could have only the vaguest grasp of the meaning of a number like 1,000,000,017. A number system is not just a collection of numbers but a collection of numbers together with rules for how to do arithmetic. Another way to summarize the abstract approach is: think about the rules rather than the numbers themselves. The numbers, from this point of view, are tokens in a sort of game (or perhaps one should call them counters).

To get some idea of what the rules are, let us consider a simple arithmetical question: what does one do if one wants to become convinced that $38 \times 263 = 9994$? Most people would probably check it on a calculator, but if this was for some reason not possible, then they might reason as follows.

$$\begin{aligned} 38 \times 263 &= 30 \times 263 + 8 \times 263 \\ &= 30 \times 200 + 30 \times 60 + 30 \times 3 + 8 \times 200 + 8 \times 60 + 8 \times 3 \\ &= 6000 + 1800 + 90 + 1600 + 480 + 24 \\ &= 9400 + 570 + 24 \\ &= 9994 \end{aligned}$$

Why, though, do these steps seem so obviously correct? For example, why does one instantly believe that $30 \times 200 = 6000$? The definition of 30 is 3×10 and the definition of 200 is $2 \times (10 \times 10)$, so we can say with total confidence that $30 \times 200 = (3 \times 10) \times (2 \times (10 \times 10))$. But why is this 6000?

Normally, nobody would bother to ask this question, but to someone who did, we might say,

$$(3 \times 10) \times (2 \times (10 \times 10)) = (3 \times 2) \times (10 \times 10 \times 10) = 6 \times 1000 = 6000$$

Without really thinking about it, we would be using two familiar facts about multiplying: that if you multiply two numbers together, it doesn't matter which order you put them in, and that if you multiply more than two numbers together, then it makes no difference how you bracket them. For example, $7 \times 8 = 8 \times 7$ and

$(31 \times 34) \times 35 = 31 \times (34 \times 35)$. Notice that the intermediate calculations involved in the second of these two examples are definitely affected by the bracketing – but one knows that the final answer will be the same.

These two rules are called the *commutative* and *associative* laws for multiplication. Let me now list a few rules, including these two, that we commonly use when adding and multiplying.

A1 The commutative law for addition: $a + b = b + a$ for any two numbers a and b.
A2 The associative law for addition: $a + (b + c) = (a + b) + c$ for any three numbers a, b, and c.
M1 The commutative law for multiplication: $ab = ba$ for any two numbers a and b.
M2 The associative law for multiplication: $a(bc) = (ab)c$ for any three numbers a, b, and c.
M3 1 is a multiplicative identity: $1a = a$ for any number a.
D The distributive law: $(a + b)c = ac + bc$ for any three numbers a, b, and c.

I list these rules not because I want to persuade you that they are interesting on their own, but to draw attention to the role they play in our thinking, even about quite simple mathematical statements. Our confidence that $2 \times 3 = 6$ is probably based on a picture such as this.

* * *

* * *

On the other hand, a direct approach is out of the question if we want to show that $38 \times 263 = 9994$, so we think about this more complicated fact in an entirely different way, using the commutative, associative, and distributive laws. If we have obeyed these rules, then we believe the result. What is more, we believe it even if we have absolutely no visual sense of what 9994 objects would look like.

Zero

Historically, the idea of the number zero developed later than that of the positive integers. It has seemed to many people to be a mysterious and paradoxical concept, inspiring questions such as, 'How can something exist and yet be nothing?' From the abstract point of view, however, zero is very straightforward – it is just a new token introduced into our number system with the following special property.

A3 0 is an additive identity: $0 + a = a$ for any number a.

That is all you need to know about 0. Not what it means – just a little rule that tells you what it does.

What about other properties of the number 0, such as the fact that 0 times any number is 0? I did not list this rule, because it turns out that it can be deduced from property **A3** and our earlier rules. Here, for example, is how to show that $0 \times 2 = 0$, where 2 is defined to be the number $1 + 1$. First, rule **M1** tells us that $0 \times 2 = 2 \times 0$. Next, rule **D** tells us that $(1 + 1) \times 0 = 1 \times 0 + 1 \times 0$. But $1 \times 0 = 0$ by rule **M3**, so this equals $0 + 0$. Rule **A3** implies that $0 + 0 = 0$, and the argument is finished.

An alternative, non-abstract argument might be something like this: '0×2 means *add up no twos*, and if you do that you are left with nothing, that is, 0.' But this way of thinking makes it hard to answer questions such as the one asked by my son John (when six): how can nought times nought be nought, since nought times nought means that you have *no* noughts? A good answer, though not one that was suitable at the time, is that it can be deduced from the rules as follows. (After each step, I list the rule I am using.)

$$\begin{aligned}
0 &= 1 \times 0 & &\mathbf{M3}\\
&= (0+1) \times 0 & &\mathbf{A3}\\
&= 0 \times 0 + 1 \times 0 & &\mathbf{D}\\
&= 0 \times 0 + 0 & &\mathbf{M3}\\
&= 0 + 0 \times 0 & &\mathbf{A1}\\
&= 0 \times 0 & &\mathbf{A3}
\end{aligned}$$

Why am I giving these long-winded proofs of very elementary facts? Again, it is not because I find the proofs mathematically interesting, but rather because I wish to show what it means to justify arithmetical statements abstractly (by using a few simple rules and not worrying what numbers actually are) rather than concretely (by reflecting on what the statements mean). It is of course very useful to associate meanings and mental pictures with mathematical objects, but, as we shall see many times in this book, often these associations are not enough to tell us what to do in new and unfamiliar contexts. Then the abstract method becomes indispensable.

Negative numbers and fractions

As anybody with experience of teaching mathematics to small children knows, there is something indirect about subtraction and division that makes them harder to understand than addition and multiplication. To explain subtraction, one can of course use the notion of taking away, asking questions such as, 'How many oranges will be left if you start with five and eat two of them?'. However, that is not always the best way to think about it. For example, if one has to subtract 98 from 100, then it is better to think not about taking 98 away from 100, but about what one has to add to 98 to make 100. Then, what one is effectively doing is solving the equation $98 + x = 100$, although of course it is unusual for the letter x actually to cross one's mind during the calculation. Similarly, there are two ways of thinking about division. To explain the meaning of 50 divided by 10, one can either ask, 'If fifty objects are split into ten equal groups, then how many will be in each group?' or ask, 'If fifty objects are split into groups of ten, then how many groups will there be?'. The second approach is equivalent to the question, 'What must ten be multiplied by to make fifty, which in turn is equivalent to solving the equation $10x = 50$.

A further difficulty with explaining subtraction and division to children is that they are not always possible. For example, you cannot take ten oranges away from a bowl of seven, and three children cannot share eleven marbles equally. However, that does not stop adults subtracting 10 from 7 or dividing 11 by 3, obtaining the answers −3 and 11/3 respectively. The question then arises: do the numbers −3 and 11/3 actually exist, and if so what are they?

From the abstract point of view, we can deal with these questions as we dealt with similar questions about zero: by forgetting about them. All we need to know about −3 is that when you add 3 to it you get 0, and all we need to know about 11/3 is that when you multiply it by 3 you get 11. Those are the rules, and, in conjunction with earlier rules, they allow us to do arithmetic in a larger number system. Why should we wish to extend our number system in this way? Because it gives us a model in which equations like $x + a = b$ and $ax = b$ can be solved, whatever the values of a and b, except that a should not be 0 in the second equation. To put this another way, it gives us a model where subtraction and division are always possible, as long as one does not try to divide by 0. (The issue of division by 0 will be discussed later in the chapter.)

As it happens, we need only two more rules to extend our number system in this way: one that gives us negative numbers and one that gives us fractions, or *rational* numbers as they are customarily known.

A4 Additive inverses: for every number a there is a number b such that $a + b = 0$.

M4 Multiplicative inverses: for every number a apart from 0 there is a number c such that $ac = 1$.

Armed with these rules, we can think of $-a$ and $1/a$ as notation for the numbers b in **A4** and c in **M4** respectively. As for a more general expression like p/q, it stands for p multiplied by $1/q$.

The rules **A4** and **M4** imply two further rules, known as cancellation laws.

A5 The cancellation law for addition: if a, b, and c are any three numbers and $a + b = a + c$, then $b = c$.
M5 The cancellation law for multiplication: if a, b, and c are any three numbers, a is not 0 and $ab = ac$, then $b = c$.

The first of these is proved by adding $-a$ to both sides, and the second by multiplying both sides by $1/a$, just as you would expect. Note the different status of **A5** and **M5** from the rules that have gone before – they are *consequences* of the earlier rules, rather than rules we simply introduce to make a good game.

If one is asked to add two fractions, such as 2/5 and 3/7, then the usual method is to give them a common denominator, as follows:

$$\frac{2}{5} + \frac{3}{7} = \frac{14}{35} + \frac{15}{35} = \frac{29}{35}$$

This technique, and others like it, can be justified using our new rules. For example,

$$35 \times \frac{14}{35} = 35 \times \left(14 \times \frac{1}{35}\right) = (35 \times 14) \times \frac{1}{35} = (14 \times 35) \times \frac{1}{35}$$
$$= 14 \times \left(35 \times \frac{1}{35}\right) = 14 \times 1 = 14,$$

and

$$35 \times \frac{2}{5} = (5 \times 7) \times \left(2 \times \frac{1}{5}\right) = (7 \times 5) \times \left(\frac{1}{5} \times 2\right) = \left(7 \times \left(5 \times \frac{1}{5}\right)\right) \times 2$$
$$= (7 \times 1) \times 2 = 7 \times 2 = 14.$$

Hence, by rule **M5**, 2/5 and 14/35 are equal, as we assumed in the calculation.

Similarly, one can justify familiar facts about negative numbers. I leave it to the reader to deduce from the rules that $(-1) \times (-1) = 1$ – the deduction is fairly similar to the proof that $0 \times 0 = 0$.

Why does it seem to many people that negative numbers are less real than positive ones? Probably it is because counting small groups of objects is a fundamental human activity, and when we do it we do not use negative numbers. But all this means is that the natural number system, thought of as a model, is useful in certain circumstances where an enlarged number system is not. If we want to think about temperatures, or dates, or bank accounts, then negative numbers *do* become useful. As long as the extended number system is logically consistent, which it is, there is no harm in using it as a model.

It may seem strange to call the natural number system a model. Don't we *actually count*, with no particular idealization involved? Yes we do, but this procedure is not always appropriate, or even possible. There is nothing wrong with the number 139484027593649864923498 from a mathematical point of view, but if we can't even count votes in Florida, it is inconceivable that we might ever be sure that we had a collection of 139484027593649864923498 objects. If you take two piles of leaves and add them to a third, the result is not three piles of leaves but one large pile. And if you have just watched a rainstorm, then, as Wittgenstein said, 'The proper answer to the question, "How many drops did you see?", is *many*, not that there was a number but you don't know how many.'

Real and complex numbers

The real number system consists of all numbers that can be represented by infinite decimals. This concept is more sophisticated than it seems, for reasons that will be explained in Chapter 4. For now, let me say that the reason for extending our number system from rational to real numbers is similar to the reason for introducing negative numbers and fractions: they allow us to solve equations that we could not otherwise solve.

The most famous example of such an equation is $x^2 = 2$. In the sixth century BC it was discovered by the school of Pythagoras that $\sqrt{2}$ is irrational, which means that it cannot be represented by a fraction. (A proof of this will be given in the next chapter.) This discovery caused dismay when it was made, but now we cheerfully accept that we must extend our number system if we want to model things like the length of the diagonal of a square. Once again, the abstract method makes our task very easy. We introduce a new symbol, $\sqrt{2}$, and have one rule that tells us what to do with it: it squares to 2.

If you are well trained, then you will object to what I have just said on the grounds that the rule does not distinguish between $\sqrt{2}$ and $-\sqrt{2}$. One way of dealing with this is to introduce a new concept into our number system, that of *order*. It is often useful to talk of one number being bigger than another, and if we allow ourselves to do that then we can pick out $\sqrt{2}$ by the additional property that $\sqrt{2}$ is greater than 0. But even without this property we can do calculations such as

$$\frac{1}{\sqrt{2}-1} = \frac{\sqrt{2}+1}{(\sqrt{2}-1)(\sqrt{2}+1)} = \frac{\sqrt{2}+1}{(\sqrt{2})^2 - \sqrt{2} + \sqrt{2} - 1} = \frac{\sqrt{2}+1}{2-1} = \sqrt{2}+1,$$

and there is actually an advantage in *not* distinguishing between $\sqrt{2}$ and $-\sqrt{2}$, which is that then we know that the above calculation is valid for both numbers.

Historical suspicion of the abstract method has left its traces in the words used to describe the new numbers that arose each time the number system was enlarged, words like 'negative' and 'irrational'. But far harder to swallow than these were the 'imaginary', or 'complex', numbers, that is, numbers of the form $a + bi$, where a and b are real numbers and i is the square root of -1.

From a concrete point of view, one can quickly dismiss the square root of -1: since the square of any number is positive, -1 does not have a square root, and that is the end of the story. However, this objection carries no force if one adopts an abstract attitude. Why not simply continue to extend our number system, by introducing a solution to the equation $x^2 = -1$ and calling it i? Why should this be more objectionable than our earlier introduction of $\sqrt{2}$?

One answer might be that $\sqrt{2}$ has a decimal expansion which can (in principle) be calculated to any desired accuracy, while nothing equivalent can be said about i. But all this says is something we already know, namely that i is not a real number – just as $\sqrt{2}$ is not a rational number. It does not stop us extending our number system to one in which we can do calculations such as

$$\frac{1}{i-1} = \frac{i+1}{(i-1)(i+1)} = \frac{i+1}{i^2 - i + i - 1} = \frac{i+1}{-1-1} = -\frac{1}{2}(i+1)$$

The main difference between i and $\sqrt{2}$ is that in the case of i we are *forced* to think abstractly, whereas there is always the option with $\sqrt{2}$ of using a concrete representation such as 1.4142 . . . or the length of the diagonal of a unit square. To see why i has no such representation, ask yourself the following question: which of the two square roots of -1 is i and which is $-i$? The question does not make sense because the *only* defining property of i is that it squares to -1. Since $-i$ has the same property, any true sentence about i remains true if one replaces it with the corresponding sentence about $-i$. It is difficult, once one has grasped this, to have any respect for the view that i might denote an independently existing Platonic object.

There is a parallel here with a well-known philosophical conundrum. Might it be that when you perceive the colour red your sensation is what I experience when I perceive green, and vice versa? Some philosophers take this question seriously and define 'qualia' to be the absolute intrinsic experiences we have when, for example, we see colours. Others do not believe in qualia. For them, a word like 'green' is defined more abstractly by its role in a linguistic system, that is, by its relationships with concepts like 'grass', 'red', and so on. It is impossible to deduce somebody's position on this issue from the way they talk about colour, except during philosophical debates. Similarly, all that matters in practice about numbers and other mathematical objects is the rules they obey.

If we introduced i in order to have a solution to the equation $x^2 = -1$, then what about other, similar equations such as $x^4 = -3$, or $2x^6 + 3x + 17 = 0$? Remarkably, it turns out that every such equation can be solved within the complex number system. In other words, we make the small investment of accepting the number i, and are then repaid many times over. This fact, which has a complicated history but is usually attributed to Gauss, is known as the fundamental theorem of algebra and it provides very convincing evidence that there is something natural about i. It may be impossible to imagine a basket of i apples, a car journey that lasts i hours, or a bank account with an overdraft of i pounds, but the complex number system has become indispensable to mathematicians, and to scientists and engineers as well – the theory of quantum mechanics, for example, depends heavily on complex numbers. It provides one of the best illustrations of a general principle: if an abstract mathematical construction is sufficiently natural, then it will almost certainly find a use as a model.

A first look at infinity

Once one has learned to think abstractly, it can be exhilarating, a bit like suddenly being able to ride a bicycle without having to worry about keeping one's balance. However, I do not wish to give the impression that the abstract method is like a licence to print money. It is interesting to contrast the introduction of i into the number system with what happens when one tries to introduce the number infinity. At first there seems to be nothing stopping us: infinity should mean something like 1 divided by 0, so why not let ∞ be an abstract symbol and regard it as a solution to the equation $0x = 1$?

The trouble with this idea emerges as soon as one tries to do arithmetic. Here, for example, is a simple consequence of **M2**, the associative law for multiplication, and the fact that $0 \times 2 = 0$.

$$1 = \infty \times 0 = \infty \times (0 \times 2) = (\infty \times 0) \times 2 = 1 \times 2 = 2$$

What this shows is that the existence of a solution to the equation $0x = 1$ leads to an *inconsistency*. Does that mean that infinity does not exist? No, it simply means that no natural notion of infinity is compatible with the laws of arithmetic. It is sometimes useful to extend the number system to include the symbol ∞, accepting that in the enlarged system these laws are not always valid. Usually, however, one prefers to keep the laws and do without infinity.

Raising numbers to negative and fractional powers

One of the greatest virtues of the abstract method is that it allows us to make sense of familiar concepts in unfamiliar situations. The phrase 'make sense' is quite appropriate, since that is just what we do, rather than discovering some pre-existing sense. A simple example of this is the way we extend the concept of raising a number to a power.

If n is a positive integer, then a^n means the result of multiplying n as together. For example, $5^3 = 5 \times 5 \times 5 = 125$ and $2^5 = 2 \times 2 \times 2 \times 2 \times 2 = 32$. But with this as a definition, it is not easy to interpret an expression such as $2^{3/2}$, since you cannot take one and a half twos and multiply them together. What is the abstract method for dealing with a problem like this? Once again, it is not to look for intrinsic meaning – in this case of expressions like a^n – but to think about rules.

Two elementary rules about raising numbers to powers are the following.

E1 $a^1 = a$ for any real number a.
E2 $a^{m+n} = a^m \times a^n$ for any real number a and any pair of natural numbers m and n.

For example, $2^5 = 2^3 \times 2^2$ since 2^5 means $2 \times 2 \times 2 \times 2 \times 2$ and $2^3 \times 2^2$ means $(2 \times 2 \times 2) \times (2 \times 2)$. These are the same number because multiplication is associative.

From these two rules we can quickly recover the facts we already know. For example, $a^2 = a^{1+1}$ which, by **E2**, is $a^1 \times a^1$. By **E1** this is $a \times a$, as it should be. But we are now in a position to do much more. Let us write x for the number $2^{3/2}$. Then $x \times x = 2^{3/2} \times 2^{3/2}$ which, by **E2**, is $2^{3/2 + 3/2} = 2^3 = 8$. In other words, $x^2 = 8$. This does not quite determine x, since 8 has two square roots, so it is customary to adopt the following convention.

E3 If $a > 0$ and b is a real number, then a^b is positive.

Using **E3** as well, we find that $2^{3/2}$ is the positive square root of 8.

This is not a discovery of the 'true value' of $2^{3/2}$. However, neither is the interpretation we have given to the expression $2^{3/2}$ arbitrary – it is the only possibility if we want to preserve rules **E1**, **E2**, and **E3**.

A similar argument allows us to interpret a^0, at least when a is not 0. By **E1** and **E2** we know that $a = a^1 = a^{1+0} = a^1 \times a^0 = a \times a^0$. The cancellation law **M5** then implies that $a^0 = 1$, whatever the value of a. As for negative powers, if we know the value of a^b, then $1 = a^0 = a^{b+(-b)} = a^b \times a^{-b}$, from which it follows that $a^{-b} = 1/a^b$. The number $2^{-3/2}$, for example, is $1/\sqrt{8}$.

Another concept that becomes much simpler when viewed abstractly is that of a logarithm. I will not have much to say about logarithms in this book, but if they worry you, then you may be reassured to learn that all you need to know in order to use them are the following three rules. (If you want logarithms to base e instead of 10, then just replace 10 by e in **L1**.)

L1 $\log(10) = 1$.
L2 $\log(xy) = \log(x) + \log(y)$.
L3 If $x < y$ then $\log(x) < \log(y)$.

For example, to see that $\log(30)$ is less than $3/2$, note that

$\log(1000) = \log(10) + \log(100) = \log(10) + \log(10) + \log(10) = 3$,

by **L1** and **L2**. But $2\log(30) = \log(30) + \log(30) = \log(900)$, by **L2**, and $\log(900) < \log(1000)$, by **L3**. Hence, $2\log(30) < 3$, so that $\log(30) < 3/2$.

I shall discuss many concepts, later in the book, of a similar nature to these. They are puzzling if you try to understand them concretely, but they lose their mystery when you relax, stop worrying about what they *are*, and use the abstract method.

Chapter 3
Proofs

The diagram below shows five circles, the first with one point on its boundary, the second with two, and so on. All the possible lines joining these boundary points have also been drawn, and these lines divide the circles into regions. If one counts the regions in each circle one obtains the sequence 1,2,4,8,16. This sequence is instantly recognizable: it seems that the number of regions doubles each time

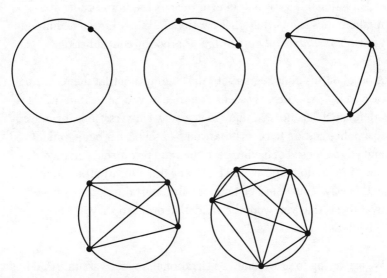

9. Dividing a circle into regions

a new point is added to the boundary, so that n points define 2^{n-1} regions, at least if no three lines meet at a point.

Mathematicians, however, are rarely satisfied with the phrase 'it seems that'. Instead, they demand a *proof*, that is, an argument that puts a statement beyond all possible doubt. What, though, does this mean? Although one can often establish a statement as true beyond all *reasonable* doubt, surely it is going a bit far to claim that an argument leaves no room for doubt whatsoever. Historians can provide many examples of statements, some of them mathematical, that were once thought to be beyond doubt, but which have since been shown to be incorrect. Why should the theorems of present-day mathematics be any different? I shall answer this question by giving a few examples of proofs and drawing from them some general conclusions.

The irrationality of the square root of two

As I mentioned in the last chapter, a number is called rational if it can be written as a fraction p/q, where p and q are whole numbers, and irrational if it cannot. One of the most famous proofs in mathematics shows that $\sqrt{2}$ is irrational. It illustrates a technique known as *reductio ad absurdum*, or proof by contradiction.

A proof of this kind begins with the assumption that the result to be proved is *false*. This may seem a strange way to argue, but in fact we often use the same technique in everyday conversation. If you went to a police station to report that you had seen a car being vandalized and were accused of having been the vandal yourself, you might well say, 'If *I* had done it, I would hardly be drawing attention to myself in this way.' You would be temporarily entertaining the (untrue) hypothesis that you were the vandal, in order to show how ridiculous it was.

We are trying to prove that $\sqrt{2}$ is irrational, so let us assume that it is *rational* and try to show that this assumption leads to absurd consequences. I shall write the argument out as a sequence of steps, giving more detail than many readers are likely to need.

1. If $\sqrt{2}$ is rational, then we can find whole numbers p and q such that $\sqrt{2} = p/q$ (by the definition of 'rational').
2. Any fraction p/q is equal to some fraction r/s where r and s are not both even. (Just keep dividing the top and bottom of the fraction by 2 until at least one of them becomes odd. For example, the fraction 1412/1000 equals 706/500 equals 353/250.)
3. Therefore, if $\sqrt{2}$ is rational, then we can find whole numbers r and s, not both even, such that $\sqrt{2} = r/s$.
4. If $\sqrt{2} = r/s$, then $2 = r^2/s^2$ (squaring both sides of the equation).
5. If $2 = r^2/s^2$, then $2s^2 = r^2$ (multiplying both sides by s^2).
6. If $2s^2 = r^2$, then r^2 is even, which means that r must be even.
7. If r is even, then $r = 2t$ for some whole number t (by the definition of 'even').
8. If $2s^2 = r^2$ and $r = 2t$, then $2s^2 = (2t)^2 = 4t^2$, from which it follows that $s^2 = 2t^2$ (dividing both sides by 2).
9. If $s^2 = 2t^2$, then s^2 is even, which means that s is even.
10. Under the assumption that $\sqrt{2}$ is rational, we have shown that $\sqrt{2} = r/s$, with r and s not both even (step 3). We have then shown that r is even (step 6) and that s is even (step 9). This is a clear contradiction. Since the assumption that $\sqrt{2}$ is rational has consequences that are clearly false, the assumption itself must be false. Therefore, $\sqrt{2}$ is irrational.

I have tried to make each of the above steps so obviously valid that the conclusion of the argument is undeniable. However, have I really left *no* room for doubt? If somebody offered you the chance of ten thousand pounds, on condition that you would have to forfeit your life if two positive whole numbers p and q were ever discovered such that $p^2 = 2q^2$, then would you accept the offer? If so, would you be in the slightest bit worried?

Step 6 contains the assertion that if r^2 is even, then r must also be even. This seems pretty obvious (an odd number times an odd number is odd) but perhaps it could do with more justification if we are trying to establish *with absolute certainty* that $\sqrt{2}$ is irrational. Let us split it into five further substeps:

6a. r is a whole number and r^2 is even. We would like to show that r must also be even. Let us assume that r is odd and seek a contradiction.

6b. Since r is odd, there is a whole number t such that $r = 2t + 1$.

6c. It follows that $r^2 = (2t + 1)^2 = 4t^2 + 4t + 1$.

6d. But $4t^2 + 4t + 1 = 2(2t^2 + 2t) + 1$, which is odd, contradicting the fact that r^2 is even.

6e. Therefore, r is even.

Does this now make step 6 completely watertight? Perhaps not, because substep 6b needs to be justified. After all, the definition of an odd number is simply a whole number that is not a multiple of two. Why should every whole number be either a multiple of two or one more than a multiple of two? Here is an argument that establishes this.

6b1. Let us call a whole number r *good* if it is either a multiple of two or one more than a multiple of two. If r is good, then $r = 2s$ or $r = 2s + 1$, where s is also a whole number. If $r = 2s$ then $r + 1 = 2s + 1$, and if $r = 2s + 1$, then $r + 1 = 2s + 2 = 2(s + 1)$. Either way, $r + 1$ is also good.

6b2. 1 is good, since $0 = 0 \times 2$ is a multiple of 2 and $1 = 0 + 1$.

6b3. Applying step 6b1 repeatedly, we can deduce that 2 is good, then that 3 is good, then that 4 is good, and so on.

6b4. Therefore, every positive whole number is good, as we were trying to prove.

Have we now finished? Perhaps the shakiest step this time is

6b4, because of the rather vague words 'and so on' from the previous step. Step 6b3 shows us how to demonstrate, for any *given* positive whole number n, that it is good. The trouble is that in the course of the argument we will have to count from 1 to n, which, if n is large, will take a very long time. The situation is even worse if we are trying to show that *every* positive whole number is good. Then it seems that the argument will never end.

On the other hand, given that steps 6b1 to 6b3 genuinely and unambiguously provide us with a method for showing that any individual n is good (provided that we have time), this objection seems unreasonable. So unreasonable, in fact, that mathematicians adopt the following principle as an axiom.

> Suppose that for every positive integer n there is associated a statement $S(n)$. (In our example, $S(n)$ stands for the statement 'n is good'.) If $S(1)$ is true, and if the truth of $S(n)$ always implies the truth of $S(n + 1)$, then $S(n)$ is true for every n.

This is known as the principle of mathematical induction, or just induction to those who are used to it. Put less formally, it says that if you have an infinite list of statements that you wish to prove, then one way to do it is to show that the first one is true and that each one implies the next.

As the last few paragraphs illustrate, the steps of a mathematical argument can be broken down into smaller and therefore more clearly valid substeps. These steps can then be broken down into subsubsteps, and so on. A fact of fundamental importance to mathematics is that this process *eventually comes to an end*. In principle, if you go on and on splitting steps into smaller ones, you will end up with a very long argument starts with axioms that are universally accepted and proceeds to the desired conclusion by means of only the most elementary logical rules (such as 'if A is true and A implies B then B is true').

What I have just said in the last paragraph is far from obvious: in fact it was one of the great discoveries of the early 20th century, largely due to Frege, Russell, and Whitehead (see Further reading). This discovery has had a profound impact on mathematics, because it means that *any dispute about the validity of a mathematical proof can always be resolved*. In the 19th century, by contrast, there were genuine disagreements about matters of mathematical substance. For example, Georg Cantor, the father of modern set theory, invented arguments that relied on the idea that one infinite set can be 'bigger' than another. These arguments are accepted now, but caused great suspicion at the time. Today, if there is disagreement about whether a proof is correct, it is either because the proof has not been written in sufficient detail, or because not enough effort has been spent on understanding it and checking it carefully.

Actually, this does not mean that disagreements never occur. For example, it quite often happens that somebody produces a very long proof that is unclear in places and contains many small mistakes, but which is not obviously incorrect in a fundamental way. Establishing conclusively whether such an argument can be made watertight is usually extremely laborious, and there is not much reward for the labour. Even the author may prefer not to risk finding that the argument is wrong.

Nevertheless, the fact that disputes can *in principle* be resolved does make mathematics unique. There is no mathematical equivalent of astronomers who still believe in the steady-state theory of the universe, or of biologists who hold, with great conviction, very different views about how much is explained by natural selection, or of philosophers who disagree fundamentally about the relationship between consciousness and the physical world, or of economists who follow opposing schools of thought such as monetarism and neo-Keynesianism.

It is important to understand the phrase 'in principle' above. No mathematician would ever bother to write out a proof in complete detail – that is, as a deduction from basic axioms using only the most utterly obvious and easily checked steps. Even if this were feasible it would be quite unnecessary: mathematical papers are written for highly trained readers who do not need everything spelled out. However, if somebody makes an important claim and other mathematicians find it hard to follow the proof, they will ask for clarification, and the process will then begin of dividing steps of the proof into smaller, more easily understood substeps. Usually, again because the audience is highly trained, this process does not need to go on for long until either the necessary clarification has been provided or a mistake comes to light. Thus, a purported proof of a result that other mathematicians care about is almost always accepted as correct only if it *is* correct.

I have not dealt with a question that may have occurred to some readers: why should one accept the axioms proposed by mathematicians? If, for example, somebody were to object to the principle of mathematical induction, how could the objection be met? Most mathematicians would give something like the following response. First, the principle seems obviously valid to virtually everybody who understands it. Second, what matters about an axiom system is less the *truth* of the axioms than their consistency and their usefulness. What a mathematical proof actually does is show that certain conclusions, such as the irrationality of $\sqrt{2}$, follow from certain premises, such as the principle of mathematical induction. The validity of these premises is an entirely independent matter which can safely be left to philosophers.

The irrationality of the golden ratio

A common experience for people learning advanced mathematics is to come to the end of a proof and think, 'I understood how each line followed from the previous one, but somehow I am none the wiser about *why* the theorem is true, or how anybody thought of this

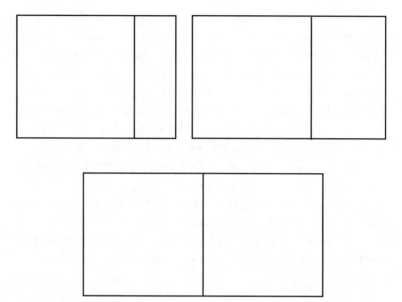

10. The existence of the golden ratio

argument'. We usually want more from a proof than a mere guarantee of correctness. We feel after reading a good proof that it provides an explanation of the theorem, that we understand something we did not understand before.

Since a large proportion of the human brain is devoted to the processing of visual data, it is not surprising that many arguments exploit our powers of visualization. To illustrate this, I shall give another proof of irrationality, this time of the so-called golden ratio. This is a number that has fascinated non-mathematicians (and to a lesser extent mathematicians) for centuries. It is the ratio of the side lengths of a rectangle with the following property: if you cut a square off it then you are left with a smaller, rotated rectangle of exactly the same shape as the original one. This is true of the second rectangle in Figure 10.

Why should such a ratio exist at all? (Mathematicians are trained to ask this sort of question.) One way to see it is to imagine a small rectangle growing out of the side of a square so that the square turns into a larger rectangle. To begin with, the small rectangle is very long and thin, while the larger one is still almost a square. If we allow the small rectangle to grow until it becomes a square itself, then the larger rectangle has become twice as long as it is wide. Thus, at first the smaller rectangle was much thinner than the larger one, and now it is fatter (relative to its size). Somewhere in between there must be a point where the two rectangles have the same shape. Figure 10 illustrates this process.

A second way of seeing that the golden ratio exists is to calculate it. If we call it x and assume that the square has side lengths 1, then the side lengths of the large rectangle are 1 and x, while the side lengths of the small one are $x - 1$ and 1. If they are the same shape, then $x = \dfrac{x}{1} = \dfrac{1}{x-1}$. Multiplying both sides by $x - 1$ we deduce that $x(x - 1) = 1$, so $x^2 - x - 1 = 0$. Solving this quadratic equation, and bearing in mind that x is not a negative number, we find that $x = \dfrac{1 + \sqrt{5}}{2}$. (If you are particularly well trained mathematically, or have taken the last chapter to heart, you may now ask why I am so confident that $\sqrt{5}$ exists. In fact, what this second argument does is to reduce a geometrical problem to an equivalent algebraic one.)

Having established that the ratio x exists, let us take a rectangle with sides of length x and 1 and consider the following process. First, cut off a square from it, leaving a smaller rectangle which, by the definition of the golden ratio, has the same shape as the original one. Now repeat this basic operation over and over again, obtaining a sequence of smaller and smaller rectangles, each of the same shape as the one before and hence each with side lengths in the golden ratio. Clearly, the process will never end. (See the first rectangle of Figure 11.)

Now let us do the same to a rectangle with side lengths in the ratio p/q, where p and q are whole numbers. This means that the

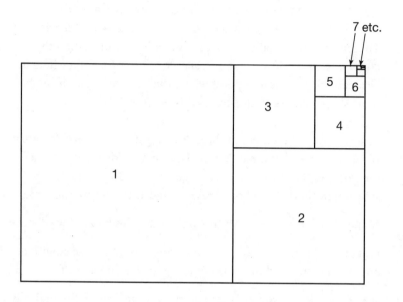

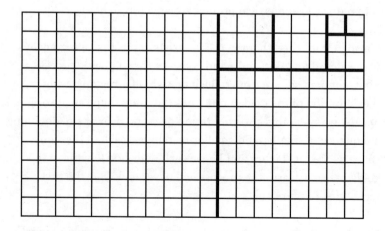

11. Cutting squares off rectangles

rectangle has the same shape as a rectangle with side lengths p and q, so it can be divided into $p \times q$ little squares, as illustrated by the second rectangle of Figure 11. What happens if we remove large squares from the end of this rectangle? If q is smaller than p, then we will remove a $q \times q$ square and end up with a $q \times (p-q)$ rectangle. We can then remove a further square, and so on. Can the process go on for ever? No, because each time we cut off a square we remove a whole number of the little squares, and we cannot possibly do this more than $p \times q$ times because there were only $p \times q$ little squares to start with.

We have shown the following two facts.

1. If the ratio of the sides of the rectangle is the golden ratio, then one can continue cutting off squares for ever.
2. If the ratio of the sides of the rectangle is p/q for some pair of whole numbers p and q, then one cannot continue cutting off squares for ever.

It follows that the ratio p/q is not the golden ratio, whatever the values of p and q. In other words, the golden ratio is irrational.

If you think very hard about the above proof, you will eventually realize that it is not as different from the proof of the irrationality of $\sqrt{2}$ as it might at first appear. Nevertheless, the way it is presented is certainly different – and for many people more appealing.

Regions of a circle

Now that I have said something about the nature of mathematical proof, let us return to the problem with which the chapter began. We have a circle with n points round its boundary, we join all pairs of these points with straight lines, and we wish to show that the number of regions bounded by these lines will be 2^{n-1}. We have already seen that this is true if n is 1, 2, 3, 4, or 5. In order to prove the statement in general, we would very much like to find a convincing *reason* for the number of regions to double each time a new point is added to the boundary. What could such a reason be?

Nothing immediately springs to mind, so one way of getting started might be to study the diagrams of the divided-up circles and see whether we notice some pattern that can be generalized. For example, three points round the boundary produce three outer regions and one central one. With four points, there are four outer regions and four inner ones. With five points, there is a central pentagon, with five triangles pointing out of it, five triangles slotting into the resulting star and making it back into a pentagon, and finally five outer regions. It therefore seems natural to think of 4 as 3 + 1, of 8 as 4 + 4, and of 16 as 5 + 5 + 5 + 1.

Does this help? We do not seem to have enough examples for a clear pattern to emerge, so let us try drawing the regions that result from six points placed round the boundary. The result appears in Figure 12. Now there are six outer regions. Each of these is next to a triangular region that points inwards. Between two neighbouring regions of this kind are two smaller triangular regions. So far we have 6 + 6 + 12 = 24 regions and have yet to count the regions inside the central hexagon. These split into three pentagons, three quadrilaterals, and one central triangle. It therefore seems natural to think of the number of regions as 6 + 6 + 12 + 3 + 3 + 1.

Something seems to be wrong, though, because this gives us 31. Have we made a mistake? As it happens, no: the correct sequence begins 1, 2, 4, 8, 16, 31, 57, 99, 163. In fact, with a little further reflection one can see that the number of regions *could not possibly* double every time. For a start, it is worrying that the number of regions defined when there are 0 points round the boundary is 1 rather than 1/2, which is what it would have to be if it doubled when the first point was put in. Though anomalies of this kind sometimes happen with zero, most mathematicians would find this particular one troubling. However, a more serious problem is that if n is a

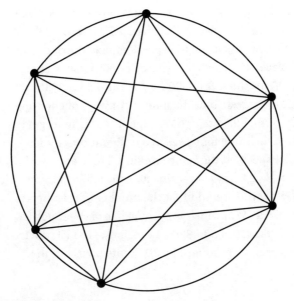

12. Regions of a circle

fairly large number, then 2^{n-1} is quite obviously too big. For example, 2^{n-1} is 524,288 when $n = 20$ and 536,870,912 when $n = 30$. Is it remotely plausible that 30 points round the edge of a circle would define over five hundred million different regions? Surely not. Imagine drawing a large circle in a field, pushing thirty pegs into it at irregularly spaced intervals, and then joining them with very thin string. The number of resulting regions would certainly be quite large, but not unimaginably so. If the circle had a diameter of ten metres and was divided into five hundred million regions, then there would have to be, on average, over six hundred regions per square centimetre. The circle would have to be thick with string, but with only thirty points round the boundary it clearly wouldn't be.

As I said earlier, mathematicians are wary of words like 'clearly'. However, in this instance our intuition can be backed up by a solid argument, which can be summarized as follows. If the circle is divided into a vast number of polygonal regions, then these regions must have, between them, a vast number of corners. Each corner is a point where two pieces of string cross, and to each such crossing one can associate four pegs, namely the ones where the relevant pieces of string end. There are 30 possible choices for the first peg, 29 for the second, 28 for the third, and 27 for the fourth. This suggests that the number of ways of choosing the four pegs is $30 \times 29 \times 28 \times 27 = 657720$, but that is to forget that if we had chosen the same four pegs in a different order, then we would have specified the same crossing. There are $4 \times 3 \times 2 \times 1 = 24$ ways of putting any given four pegs in order, and if we allow for this we find that the number of crossings is $657720/24 = 27405$, which is nothing like vast enough to be the number of corners of 536,870,912 regions. (In fact, the true number of regions produced by 30 points turns out to be 27,841.)

This cautionary tale contains many important lessons about the justification of mathematical statements. The most obvious one is that if you do not take care to prove what you say, then you run the risk of saying something that is wrong. A more positive moral is that if you do try to prove statements, then you will understand them in a completely different and much more interesting way.

Pythagoras' theorem

The famous theorem of Pythagoras states that if a right-angled triangle has sides of length a, b, and c, where c is the length of the hypotenuse (the side opposite the right angle), then $a^2 + b^2 = c^2$. It has several proofs, but one stands out as particularly short and easy to understand. Indeed, it needs little more than the following two diagrams.

In Figure 13, the squares that I have labelled A, B, and C have sides of length a, b, and c respectively, and therefore areas a^2, b^2, and c^2. Since moving the four triangles does not change their area or cause them to overlap, the area of the part of the big square that they do

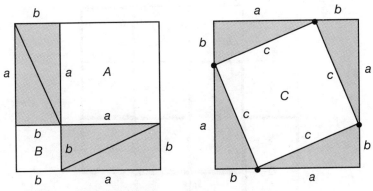

13. A short proof of Pythagoras' theorem

not cover is the same in both diagrams. But on the left this area is $a^2 + b^2$ and on the right it is c^2.

Tiling a square grid with the corners removed

Here is a well-known brainteaser. Take an eight-by-eight grid of squares and remove the squares from two opposite corners. Can you cover the remainder of the grid with domino-shaped tiles, each of which exactly covers two adjacent squares? My illustration (Figure 14) shows that you cannot if the eight-by-eight grid is replaced by a four-by-four one. Suppose you decide to place a tile in the position I have marked A. It is easy to see that you are then forced to put tiles in positions B, C, D, and E, leaving a square that cannot be covered. Since the top right-hand corner must be covered somehow, and the only other way of doing it leads to similar problems (by the symmetry of the situation), tiling the whole shape is impossible.

If we replace four by five, then tiling the grid is still impossible, for the simple reason that each tile covers two squares and there are 23 squares to cover – an odd number. However, $8^2 - 2 = 62$ is an even number, so we cannot use this argument for an eight-by-eight grid. On the other hand, if you try to find a proof similar to the one I used for a four-by-four grid, you will soon give up, as the number of

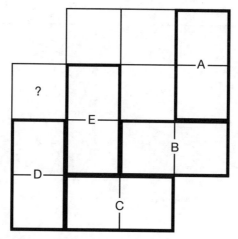

14. Tiling a square grid with the corners removed

possibilities you have to consider is very large. So how should you approach the problem? If you have not come across this question, I would urge you to try to solve it before reading on, or to skip the next paragraph, because if you manage to solve it you will have a good idea of the pleasures of mathematics.

For those who have disregarded my advice, and experience suggests that they will be in the majority, here is one word which is almost the whole proof: chess. A chessboard is an eight-by-eight grid, with its squares coloured alternately black and white (quite unnecessarily as far as the game is concerned, but making it easier to take in visually). Two opposite corner squares will have the same colour. If they are black, as they may as well be, then once they are removed the depleted chessboard has 32 white squares and 30 black ones. Each domino covers exactly one square of each colour, so once you have put down 30 dominoes, you will be left, no matter how you did it, with two white squares, and these you will be unable to cover.

This short argument illustrates very well how a proof can offer more than just a guarantee that a statement is true. For example, we now have two proofs that the four-by-four grid with two opposite corners removed cannot be tiled. One is the proof I gave and the other is the four-by-four version of the chessboard argument. Both of them establish what we want, but only the second gives us anything like a *reason* for the tiling being impossible. This reason instantly tells us that tiling a ten-thousand-by-ten-thousand grid with two opposite corners removed is also impossible. By contrast, the first argument tells us only about the four-by-four case.

It is a notable feature of the second argument that it depends on a single idea, which, though unexpected, seems very natural as soon as one has understood it. It often puzzles people when mathematicians use words like 'elegant', 'beautiful', or even 'witty' to describe proofs, but an example such as this gives an idea of what they mean. Music provides a useful analogy: we may be entranced when a piece moves in an unexpected harmonic direction that later comes to seem wonderfully appropriate, or when an orchestral texture appears to be more than the sum of its parts in a way that we do not fully understand. Mathematical proofs can provide a similar pleasure with sudden revelations, unexpected yet natural ideas, and intriguing hints that there is more to be discovered. Of course, beauty in mathematics is not the same as beauty in music, but then neither is musical beauty the same as the beauty of a painting, or a poem, or a human face.

Three obvious-seeming statements that need proofs

An aspect of advanced mathematics that many find puzzling is that some of its theorems seem too obvious to need proving. Faced with such a theorem, people will often ask, 'If *that* doesn't count as obvious, then what does?' A former colleague of mine had a good answer to this question, which is that a statement is obvious if a proof instantly springs to mind. In the remainder of this chapter, I shall give three examples of statements that may appear obvious but which do not pass this test.

1. The fundamental theorem of arithmetic states that every natural number can be written in one and only one way as a product of prime numbers, give or take the order in which you write them. For example, $36 = 2 \times 2 \times 3 \times 3$, $74 = 2 \times 37$, and 101 is itself a prime number (which is thought of in this context as a 'product' of one prime only). Looking at a few small numbers like this, one rapidly becomes convinced that there will never be two *different* ways of expressing a number as a product of primes. That is the main point of the theorem and it hardly seems to need a proof.

But is it really so obvious? The numbers 7, 13, 19, 37, and 47 are all prime, so if the fundamental theorem of arithmetic is obvious then it should be obvious that $7 \times 13 \times 19$ does not equal 37×47. One can of course check that the two numbers are indeed different (one, as any mathematician will tell you, is more interesting than the other), but that doesn't show that they were *obviously* going to be different, or explain why one could not find two other products of primes, this time giving the same answer. In fact, there is no easy proof of the theorem: if a proof instantly springs to mind, then you have a very unusual mind.

2. Suppose that you tie a slip knot in a normal piece of string and then fuse the ends together, obtaining the shape illustrated in Figure 15, known to mathematicians as the trefoil knot. Is it possible to untie the knot without cutting the string? No, of course it isn't.

Why, though, are we inclined to say 'of course'? Is there an argument that immediately occurs to us? Perhaps there is – it seems as though any attempt to untie the knot will inevitably make it more tangled rather than less. However, it is difficult to convert this instinct into a valid proof. All that is genuinely obvious is that there is no *simple* way to untie the knot. What is difficult is to rule out the possibility that there is a way of untying the trefoil knot *by making it much more complicated first*. Admittedly, this seems unlikely, but phenomena of this kind do occur in mathematics, and

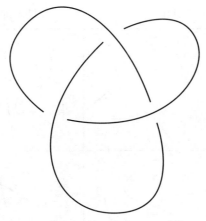

15. A trefoil knot

even in everyday life: for example, in order to tidy a room properly, as opposed to stuffing everything into cupboards, it is often necessary to begin by making it much untidier.

3. A *curve* in the plane means anything that you can draw without lifting your pen off the paper. It is called *simple* if it never crosses itself, and *closed* if it ends up where it began. Figure 16 shows what these definitions mean pictorially. The first curve illustrated, which is both simple and closed, encloses a single region of the plane, which is known as the *interior* of the curve. Clearly, every simple closed curve splits the plane into two parts, the inside and the outside (three parts if one includes the curve itself as a part).

Is this really so clear though? Yes, it certainly is if the curve is not too complicated. But what about the curve shown in Figure 17? If you choose a point somewhere near the middle, it is not altogether obvious whether it lies inside the curve or outside it. Perhaps not, you might say, but there will certainly *be* an inside and an outside, even if the complexity of the curve makes it hard to distinguish them visually.

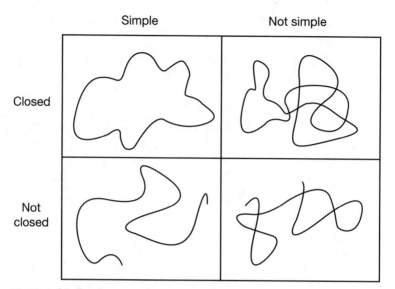

16. Four kinds of curve

How can one justify this conviction? One might attempt to distinguish the inside from the outside as follows. Assuming for a moment that the concepts of inside and outside do make sense, then every time you cross the curve, you must go from the inside to the outside or vice versa. Hence, if you want to decide whether a point P is inside or outside, all you need to do is draw a line that starts at P and ends up at some other point Q that is far enough from the curve to be quite clearly outside. If this line crosses the curve an odd number of times then P is inside; otherwise it is outside.

The trouble with this argument is that it takes various things for granted. For example, how do you know that if you draw *another* line from P, ending at a different point R, you won't get a different answer? (You won't, but this needs to be proved.) The statement that every simple closed curve has an inside and an outside is in fact a famous mathematical theorem, known as the Jordan Curve Theorem. However obvious it may seem, it needs a proof, and all

17. Is the black spot inside the curve or outside it?

known proofs of it are difficult enough to be well beyond the scope of a book like this.

Chapter 4
Limits and infinity

In the last chapter, I tried to indicate how the notion of a mathematical proof can, in principle, be completely formalized. If one starts with certain axioms, follows certain rules, and ends up with an interesting mathematical statement, then that statement will be accepted as a theorem; otherwise, it will not. This idea, of deducing more and more complicated theorems from just a few axioms, goes back to Euclid, who used just five axioms to build up large parts of geometry. (His axioms are discussed in Chapter 6.) Why, one might then ask, did it take until the 20th century for people to realize that this could be done for the whole of mathematics?

The main reason can be summed up in one word: 'infinity'. In one way or another, the concept of infinity is indispensable to mathematics, and yet it is a very hard idea to make rigorous. In this chapter I shall discuss three statements. Each one looks innocent enough at first, but turns out, on closer examination, to involve the infinite. This creates difficulties, and most of this chapter will be about how to deal with them.

1. The square root of 2 is about 1.41421356

Where is infinity involved in a simple statement like the above, which says merely that one smallish number is roughly equal to another? The answer lies in the phrase 'the square root of 2', in which it is implicitly assumed that 2 *has* a square root. If we want to understand the statement completely, this phrase forces us to ask what sort of object the square root of 2 is. And that is where infinity comes in: the square root of 2 is an infinite decimal.

Notice that there is no mention of infinity in the following closely related statement: 1.41421356 squared is close to 2. This statement is entirely finite, and yet it seems to say roughly the same thing. As we shall see later, that is important.

What does it mean to say that there is an infinite decimal which, when squared, gives 2? At school we are taught how to multiply finite decimals but not infinite ones – it is somehow just assumed that they can be added and multiplied. But how is this to be done? To see the sort of difficulty that can arise, let us consider addition first. When we add two finite decimals, such as, say, 2.3859 and 3.1405, we write one under the other and add corresponding digits, starting from the right. We begin by adding the final digits, 9 and 5, together. This gives us 14, so we write down the 4 and carry the 1. Next, we add the penultimate digits, 5 and 0, and the carried 1, obtaining 6. Continuing in this way, we reach the answer, 5.5264.

Now suppose we have two infinite decimals. We cannot start from the right, because an infinite decimal has no last digit. So how can we possibly add them together? There is one obvious answer: start from the left. However, there is a drawback in doing so. If we try it with the finite decimals 2.3859 and 3.1405, for example, we begin by adding the 2 to the 3, obtaining 5. Next, just to the right of the decimal point, we add 3 and 1 and get 4, which is, unfortunately, incorrect.

This incorrectness is inconvenient, but it is not a disaster if we keep our nerve and continue. The next two digits to be added are 8 and 4, and we can respond to them by writing down 2 as the third digit of the answer and *correcting* the second digit by changing it from 4 to 5. This process continues with our writing down 5 as the fourth digit of the answer, which will then be corrected to 6.

Notice that corrections may take place a long time after the digit has been written down. For example, if we add 1.3555555555555555573 to 2.5444444444444444452, then we begin by writing 3.8999999999999999, but that entire string of nines has to be corrected when we get to the next step, which is to add 7 to 5. Then, like a line of dominoes, the nines turn into zeros as we carry one back and back. Nevertheless, the method works, giving an answer of 3.9000000000000000025, and it enables us to give a meaning to the idea of adding two infinite decimals. It is not too hard to see that no digit will ever need to be corrected more than once, so if we have two infinite decimals, then, for example, the 53rd digit of their sum will be what we write down at the 53rd stage of the above process, or the correction of it, should a correction later become necessary.

We would like to make sense of the assertion that there is an infinite decimal whose square is 2. To do this, we must first see how this infinite decimal is generated and then understand what it means to multiply it by itself. As one might expect, multiplication of infinite decimals is more complicated than addition.

First, though, here is a natural way to generate the decimal. It has to lie between 1 and 2, because $1^2 = 1$, which is less than 2, and $2^2 = 4$, which is greater. If you work out 1.1^2, 1.2^2, 1.3^2, and so on up to 1.9^2 you find that $1.4^2 = 1.96$, which is less than 2, and $1.5^2 = 2.25$, which is greater. So $\sqrt{2}$ must lie between 1.4 and 1.5, and therefore its decimal expansion must begin 1.4. Now suppose that you have worked out in this way that the first eight digits of $\sqrt{2}$ are 1.4142135. You can then do the following calculations, which show that the next digit is 6.

$$1.41421350^2 = 1.9999998236822500$$
$$1.41421351^2 = 1.9999998519665201$$
$$1.41421352^2 = 1.9999998802507904$$
$$1.41421353^2 = 1.9999999085350609$$
$$1.41421354^2 = 1.9999999368193316$$
$$1.41421355^2 = 1.9999999651036025$$
$$1.41421356^2 = 1.9999999933878736$$
$$1.41421357^2 = 2.0000000216721449$$

Repeating this procedure, you can generate as many digits as you like. Though you will never actually finish, you do at least have an unambiguous way of defining the nth digit after the decimal point, whatever the value of n: it will be the same as the final digit of the largest decimal that squares to less than 2 and has n digits after the decimal point. For example, 1.41 is the largest decimal that squares to less than 2 with two digits after the decimal point, so the square root of two begins 1.41.

Let us call the resulting infinite decimal x. What makes us so confident that $x^2 = 2$? We might argue as follows.

$$1^2 = 1$$
$$1.4^2 = 1.96$$
$$1.41^2 = 1.9881$$
$$1.414^2 = 1.999396$$
$$1.4142^2 = 1.99996164$$
$$1.41421^2 = 1.9999899241$$
$$1.414213^2 = 1.999998409469$$
$$1.4142135^2 = 1.99999982368225$$
$$1.41421356^2 = 1.9999999933878736$$

As the above table of calculations demonstrates, the more digits we use of the decimal expansion of $\sqrt{2}$, the more nines we get after the decimal point when we multiply the number by itself. Therefore, if we use the entire infinite expansion of $\sqrt{2}$, we should get infinitely many nines, and 1.99999999 . . . (one point nine recurring) equals 2.

This argument leads to two difficulties. First, why does one point nine recurring equal two? Second, and more serious, what does it mean to 'use the entire infinite expansion'? That is what we were trying to understand in the first place.

To dispose of the first objection, we must once again set aside any Platonic instincts. It is an accepted truth of mathematics that one point nine recurring equals two, but this truth was not discovered by some process of metaphysical reasoning. Rather, it is a *convention*. However, it is by no means an arbitrary convention, because not adopting it forces one either to invent strange new objects or to abandon some of the familiar rules of arithmetic. For example, if you hold that $1.999999\ldots$ does not equal 2, then what is $2 - 1.999999\ldots$? If it is zero, then you have abandoned the useful rule that x must equal y whenever $x - y = 0$. If it is not zero, then it does not have a conventional decimal expansion (otherwise, subtract it from two and you will not get one point nine recurring but something smaller) so you are forced to invent a new object such as 'nought followed by a point, then infinitely many noughts, and *then* a one'. If you do this, then your difficulties are only just beginning. What do you get when you multiply this mysterious number by itself? Infinitely many noughts, then infinitely many noughts again, and *then* a one? What happens if you multiply it by ten instead? Do you get 'infinity minus one' noughts followed by a one? What is the decimal expansion of 1/3? Now multiply that number by 3. Is the answer 1 or $0.999999\ldots$? If you follow the usual convention, then tricky questions of this kind do not arise. (Tricky but not impossible: a coherent notion of 'infinitesimal' numbers was discovered by Abraham Robinson in the 1960s, but non-standard analysis, as his theory is called, has not become part of the mathematical mainstream.)

The second difficulty is a more genuine one, but it can be circumvented. Instead of trying to imagine what would actually happen if one applied some kind of long multiplication procedure to infinite decimals, one interprets the statement $x^2 = 2$ as meaning simply that the more digits one takes of x, the closer the square of the resulting number is to 2, just as we observed. To be more precise, suppose you insist that you want a number that, when squared, produces a number that begins 1.9999. . . . I will suggest the number 1.41421, given by the first few digits of x. Since 1.41421 is very close to 1.41422, I expect that their squares are also very close (and this can be proved quite easily). But because of how we chose x, 1.41421^2 is less than 2 and 1.41422^2 is greater than 2. It follows that both numbers are very close to 2. Just to check: $1.41421^2 = 1.9999899241$, so I have found a number with the property you wanted. If you now ask for a number that, when squared, begins

1.9999999999999999999999999999999999999999999999999 . . .,

I can use exactly the same argument, but with a few more digits of x. (It turns out that if you want n nines then $n + 1$ digits after the decimal point will always be enough.) The fact that I can do this, however many nines you want, is what is meant by saying that the infinite decimal x, when multiplied by itself, equals 2.

Notice that what we have done is to 'tame' the infinite, by interpreting a statement that involves infinity as nothing more than a colourful shorthand for a more cumbersome statement that doesn't. The neat infinite statement is 'x is an infinite decimal that squares to 2'. The translation is something like, 'There is a rule that, for any n, unambiguously determines the nth digit of x. This allows us to form arbitrarily long finite decimals, and their squares can be made as close as we like to 2 simply by choosing them long enough.'

Am I saying that the true meaning of the apparently simple statement that $x^2 = 2$ is actually very complicated? In a way I am – the statement really does have hidden complexities – but in a more important way I am not. It is hard work to define addition and multiplication of infinite decimals without mentioning infinity, and one must check that the resulting complicated definitions obey the rules set out in Chapter 2, such as the commutative and associative laws. However, once this has been done, we are free to think abstractly again. What matters about x is that it squares to two. What matters about the word 'squares' is that its meaning is based on *some* definition of multiplication that obeys the appropriate rules. It doesn't really matter what the trillionth digit of x is and it doesn't really matter that the definition of multiplication is somewhat complicated.

2. We reached a speed of 40 m.p.h. just as we passed that lamp-post

Suppose you are in an accelerating car and you have watched the speedometer move steadily from 30 m.p.h. to 50 m.p.h. It is tempting to say that just for an instant – the exact instant at which the arm of the speedometer passed 40 – the car was travelling at 40 m.p.h. Before that instant it was slower and afterwards it was faster. But what does it mean to say that the speed of a car is 40 m.p.h. just for an instant? If a car is not accelerating, then we can measure how many miles it goes in an hour, and that gives us its speed. (Alternatively, and more practically, we can see how far it goes in 30 seconds and multiply by 120.) However, this method obviously does not work with an accelerating car: if one measures how far it has gone in a certain time, all one can calculate is the *average* speed during that time, which does not tell us the speed at any given moment.

The problem would disappear if we could measure how far the car travelled during an *infinitely small* period of time, because then the speed would not have time to change. If the period of time lasted for t hours, where t was some infinitely small number, then we would measure how many miles the car travelled during those t hours, take our answer s, which would also be infinitely small of course, and divide it by t to obtain the instantaneous speed of the car.

This ridiculous fantasy leads to problems very similar to those encountered when we briefly entertained the idea that one point nine recurring might not equal two. Is t zero? If so, then quite clearly s must be as well (a car cannot travel any distance in no time at all). But one cannot divide zero by zero and obtain an unambiguous answer. On the other hand, if t is not zero, then the car accelerates during those t hours and the measurement is invalid.

The way to understand instantaneous speed is to exploit the fact that the car does not have time to accelerate *very much* if t is very small – say a hundredth of a second. Suppose for a moment that we do not try to calculate the speed exactly, but settle instead for a good estimate. Then, if our measuring devices are accurate, we can see how far the car goes in a hundredth of a second, and multiply that distance by the number of hundredths of a second in an hour, or 360,000. The answer will not be quite right, but since the car cannot accelerate much in a hundredth of a second it will give us a close approximation.

This situation is reminiscent of the fact that 1.4142135^2 is a close approximation to 2, and it allows us to avoid worrying about the infinite, or in this case infinitely small, in a very similar way. Suppose that instead of measuring how far the car went in a hundredth of a second we had measured its distance over a millionth of a second. The car would have accelerated even less during that time, so our answer would have been more accurate still. This observation gives us a way of translating the statement, 'The car is travelling at 40 m.p.h. . . . NOW!' into the following more complicated finite one: 'If you specify what margin of error I am allowed, then as long as t is a small enough number of hours (typically far less than one) I can see how many miles the car travels in t hours, divide by t and obtain an answer that will be at least as close to 40 m.p.h. as the error margin you allowed me.' For example, if t is small enough, then I can guarantee that my estimate will be between 39.99 and 40.01. If you have asked for an answer accurate to within 0.0001, then I may have to make *t* smaller, but as long as it is small enough I can give you the accuracy you want.

Once again, we are regarding a statement that involves infinity as a convenient way of expressing a more complicated statement concerning approximations. Another word, which can be more suggestive, is 'limit'. An infinite decimal is a limit of a sequence of finite decimals, and the instantaneous speed is the limit of the estimates one makes by measuring the distance travelled over shorter and shorter periods of time. Mathematicians often talk about what happens 'in the limit', or 'at infinity', but they are aware as they do so that they are not being wholly serious. If pressed to say exactly what they mean, they will start to talk about approximations instead.

3. The area of a circle of radius *r* is πr^2

The realization that the infinite can be understood in terms of the finite was one of the great triumphs of 19th-century mathematics, although its roots go back much earlier. In discussing my next example, how to calculate the area of a circle, I shall use an argument invented by Archimedes in the 3rd century BC. Before we do this calculation, though, we ought to decide what it is that we are calculating, and this is not as easy as one might think. What *is* area? Of course, it is something like the *amount of stuff* in the shape (two-dimensional stuff, that is), but how can one make this precise?

Whatever it is, it certainly seems to be easy to calculate for certain shapes. For example, if a rectangle has side lengths *a* and *b*, then its area is *ab*. Any right-angled triangle can be thought of as the result of cutting a rectangle in half down one of its diagonals, so its area is half that of the corresponding rectangle. Any triangle can be cut into two right-angled triangles, and any polygon can be split up into triangles. Therefore, it is not too difficult to work out the area of a polygon. Instead of worrying about what exactly it is that we have calculated, we can simply *define* the area of a polygon to be the result of the calculation (once we have convinced ourselves that chopping a polygon into triangles in two different ways will not give us two different answers).

Our problems begin when we start to look at shapes with curved boundaries. It is not possible to cut a circle up into a few triangles. So what are we talking about when we say that its area is πr^2?

This is another example where the abstract approach is very helpful. Let us concentrate not on what area *is*, but on what it *does*. This suggestion needs to be clarified since area doesn't seem to do all that much – surely it is just there. What I mean is that we should focus on the properties that any reasonable concept of area will have. Here are five.

Ar1 If you slide a shape about, its area does not change. (Or more formally: two congruent shapes have the same area.)

Ar2 If one shape is entirely contained in another, then the area of the first is no larger than the area of the second.

Ar3 The area of a rectangle is obtained by multiplying its two side lengths.

Ar4 If you cut a shape into a few pieces, then the areas of the pieces add up to the area of the original shape.

Ar5 If you expand a shape by a factor of 2 in every direction, then its area multiplies by 4.

If you look back, you will see that we used properties **Ar1**, **Ar3**, and **Ar4** to calculate the area of a right-angled triangle. Property **Ar2** may seem so obvious as to be hardly worth mentioning, but this is what one expects of axioms, and we shall see later that it is very useful. Property **Ar5**, though important, is not actually needed as an axiom because it can be deduced from the others.

How can we use these properties to say what we mean by the area of a circle? The message of this chapter so far is that it may be fruitful to think about *approximating* the area, rather than defining it in one go. This can be done quite easily as follows. Imagine that a shape is drawn on a piece of graph paper with a very fine grid of squares. We know the area of these squares, by property **Ar3** (since a square is a special kind of rectangle), so we could try estimating the area of the shape by counting how many of the squares lie completely within it. If, for example, the shape contains 144 squares, then the area of the shape is at least 144 times the area of each square. Notice that what we have really calculated is the area of a shape made out of the 144 squares, which is easily determined by properties **Ar3** and **Ar4**.

For the shape illustrated in Figure 18 this does not give the right answer because there are several squares that are partly in the shape

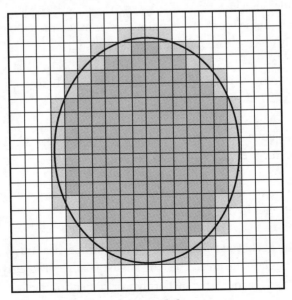

18. **Approximating the area of a curved shape**

and partly out of it, so we have not taken account of all of the area. However, there is an obvious way to improve the estimate, which is to divide each square into four smaller ones and use those instead. As before, some of the squares will be partly in and partly out, but we will now have included a little more of the shape amongst the squares that are completely inside it. In general, the finer the grid of squares, the more of the original shape we take account of in our calculation. We find (and this is not quite as obvious as it seems) that as we take finer and finer grids, with smaller and smaller squares, the results of our calculations are closer and closer to some number, just as the results of squaring better and better approximations to $\sqrt{2}$ become closer and closer to 2, and we *define* this number to be the area of the shape.

Thus, to the mathematically inclined, the statement that some shape has an area of one square yard means the following. If a certain amount of error is regarded as tolerable, then, however small it is, one can choose a sufficiently fine grid of squares, do the approximate calculation by adding up the areas of the squares inside the shape, and arrive at an answer which will differ from one square yard by less than that amount. (Somewhere in the back of one's mind, but firmly suppressed, may be the idea that 'in the limit' one could use infinitely many infinitely small squares and get the answer exactly right.)

Another way of putting it, which is perhaps clearer, is this. If a curved shape has an area of exactly 12 square centimetres, and I am required to demonstrate this using a grid of squares, then my task is impossible – I would need infinitely many of them. If, however, you give me any number *other* than 12, such as 11.9, say, then I can use a grid of squares to prove conclusively that the area of the shape is *not* that number: all I have to do is choose a grid fine enough for the area left out to be less than 0.1 square centimetres. In other words, I can do without infinity if, instead of proving that the area is 12, I settle for proving that it isn't anything else. The area of the shape is the one number I cannot disprove.

These ideas give a satisfactory definition of area, but they still leave us with a problem. How can we show that if we use the above procedure to estimate the area of a circle of radius r, then our estimates will get closer and closer to πr^2? The answer for most shapes is that one has to use the integral calculus, which I do not discuss in this book, but for the circle one can use an ingenious argument of Archimedes, as I mentioned earlier.

Figure 19 shows a circle cut into slices and an approximately rectangular shape formed by taking the slices apart and reassembling them. Because the slices are thin, the height of the rectangle is approximately the radius, r, of the circle. Again because the slices are thin, the top and bottom sides of the approximate rectangle are approximately straight lines. Since each of these sides uses half the circumference of the circle, and by the definition of π the circumference has length $2\pi r$, their lengths are approximately πr. Therefore, the area of the approximate rectangle is $r \times \pi r = \pi r^2$ – at least approximately.

Of course, it is in fact exactly πr^2 since all we have done is cut up a circle and move the pieces around, but we do not yet know this. The argument so far may have already convinced you, but it is not quite finished, as we must establish that the above approximation becomes closer and closer to πr^2 as the number of slices becomes larger and larger. Very briefly, one way to do this is to take two regular polygons, one just contained in the circle and the other just

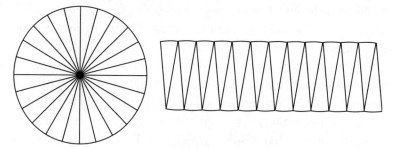

19. Archimedes' method for showing that a circle has area πr^2

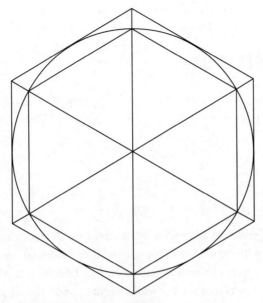

20. Approximating a circle by a polygon

containing it. Figure 20 illustrates this with hexagons. The perimeter of the inner polygon is shorter than the circumference of the circle, while that of the outer polygon is longer. The two polygons can each be cut into triangular slices and reassembled into parallelograms. A straightforward calculation shows that the smaller parallelogram has an area of less than r times half the perimeter of the inner polygon, and hence less than πr^2. Similarly, the area of the larger polygon is greater than πr^2. And yet, if the number of slices is large enough, the difference in area between the two polygons can be made as small as one likes. Since the circle always contains the smaller one and is contained in the larger, its area must be exactly πr^2.

Chapter 5
Dimension

A notable feature of advanced mathematics is that much of it is concerned with geometry in more than three dimensions. This fact is baffling to non-mathematicians: lines and curves are one-dimensional, surfaces are two-dimensional, and solid objects are three-dimensional, but how could something be four-dimensional? Once an object has height, width, and depth, it completely fills up a portion of space, and there just doesn't seem to be scope for any further dimensions. It is sometimes suggested that the fourth dimension is time, which is a good answer in certain contexts, such as special relativity, but does not help us to make sense of, say, twenty-six-dimensional or even infinite-dimensional geometry, both of which are of mathematical importance.

High-dimensional geometry is yet another example of a concept that is best understood from an abstract point of view. Rather than worrying about the *existence*, or otherwise, of twenty-six-dimensional space, let us think about its *properties*. You might wonder how it is possible to consider the properties of something without establishing first that it exists, but this worry is easily disposed of. If you leave out the words 'of something', then the question becomes: how can one consider a set of properties without first establishing that there is something that *has* those properties? But this is not difficult at all. For example, one can speculate about the character a female president of the United States would be likely to have, even though there is no guarantee that there will ever be one.

What sort of properties might we expect of twenty-six dimensional space? The most obvious one, the property that would *make* it twenty-six dimensional, is that it should take twenty-six numbers to specify a point, just as it takes two numbers in two dimensions and three in three. Another is that if you take a twenty-six dimensional shape and expand it by a factor of two in every direction, then its 'volume', assuming we can make sense of it, should multiply by 2^{26}. And so on.

Such speculations would not be very interesting if it turned out that there was something logically inconsistent about the very notion of twenty-six dimensional space. To reassure ourselves about this, we would after all like to show that it exists – which it obviously couldn't if it involved an inconsistency – but in a mathematical rather than a physical sense. What this means is that we need to define an appropriate model. It may not necessarily be a model *of* anything, but if it has all the properties we expect, it will show that those properties are consistent. As so often happens, though, the model we shall define turns out to be very useful.

How to define high-dimensional space

Defining the model is surprisingly easy to do, as soon as one has had one idea: coordinates. As I have said, a point in two dimensions can be specified using two numbers, while a point in three dimensions needs three. The usual way to do this is with Cartesian coordinates, so called because they were invented by Descartes (who maintained that he had been given the idea in a dream). In two dimensions you start with two directions at right angles to each other. For example, one might be to the right and the other directly upwards, as shown in Figure 21. Given any point in the plane, you can reach it by going a certain distance horizontally (if you go left then you regard yourself as having gone a negative distance to the right), then

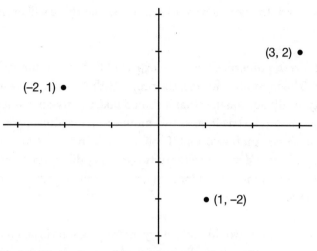

21. Three points in the Cartesian plane

turning through ninety degrees and going some other distance vertically. These two distances give you two numbers, and these numbers are the coordinates of the point you have reached. Figure 21 shows the points with coordinates (3, 2) (three to the right and two up), (–2, 1) (two to the left and one up), and (1, –2) (one to the right and two down). Exactly the same procedure works in three dimensions, that is, in space, except that you must use three directions, such as forwards, to the right, and upwards.

Now let us change our point of view very slightly. Instead of calling the two (or three) numbers *the coordinates of* a point in space, let us say that the numbers *are* the point. That is, instead of saying 'the point with coordinates (5, 3)', let us say 'the point (5, 3)'. One might regard doing this as a mere linguistic convenience, but actually it is more than that. It is replacing real, physical space with a mathematical model of space. Our mathematical model of two-dimensional space consists of pairs of real numbers (a, b). Although these pairs of numbers are not themselves points in space, we call them points because we wish to remind ourselves that that is what they represent. Similarly, we can obtain a model of three-dimensional space by taking all triples (a, b, c), and again calling them points. There is now an obvious way of defining points in, say, eight-dimensional space. They are nothing other than octuples of real numbers. For example, here are two points: $(1, 3, -1, 4, 0, 0, 6, 7)$ and $(5, \pi, -3/2, \sqrt{2}, 17, 89.93, -12, \sqrt{2} + 1)$.

I have now defined a mathematical model of sorts, but it is not yet worthy of being called a model of eight-dimensional *space*, because the word 'space' carries with it many geometrical connotations which I have not yet described in terms of the model: there is more to space than just a vast collection of single points. For example, we talk about the distance between a pair of points, and about straight lines, circles, and other geometrical shapes. What are the counterparts of these ideas in higher dimensions?

There is a general method for answering many questions of this kind. Given a familiar concept from two or three dimensions, first describe it entirely in terms of coordinates and then hope that the

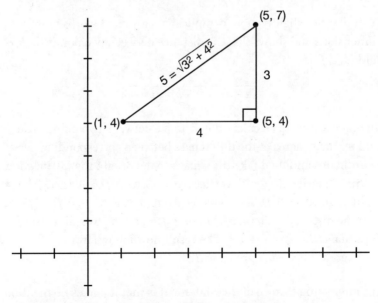

22. Calculating distances using Pythagoras' theorem

generalization to higher dimensions becomes obvious. Let us see how this works for the concept of distance.

Given two points in the plane, such as (1, 4) and (5, 7), we can calculate their distance as follows. We begin by forming a right-angled triangle with the extra point (5, 4), as illustrated in Figure 22. We then notice that the line joining the points (1, 4) to (5, 7) is the hypotenuse of this triangle, which means that its length can be calculated using Pythagoras' theorem. The lengths of the other two sides are $5 - 1 = 4$ and $7 - 4 = 3$, so the length of the hypotenuse is $\sqrt{4^2 + 3^2} = \sqrt{16 + 9} = 5$. Thus, the distance between the two points is 5. Applying this method to a general pair of points (a, b) and (c, d), we obtain a right-angled triangle with these two points at either end of the hypotenuse and two other side lengths of $|c - a|$ (this means the difference between c and a) and $|d - b|$. Pythagoras' theorem then tells us that the distance between the two points is given by the formula

$$\sqrt{(c - a)^2 + (d - b)^2}$$

A similar but slightly more complicated argument works in three dimensions and shows that the distance between two points (a, b, c) and (d, e, f) is

$$\sqrt{(d - a)^2 + (e - b)^2 + (f - c)^2}$$

In other words, to calculate the distance between two points, you add up the squares of the differences between corresponding coordinates and then take the square root. (Briefly, the justification is this. The triangle T with vertices (a, b, c), (a, b, f), and (d, e, f) has a right angle at (a, b, f). The distance from (a, b, c) to (a, b, f) is $|f - c|$ and the distance from (a, b, f) to (d, e, f) is, by the two-dimensional formula, $\sqrt{(d - a)^2 + (e - b)^2}$. The result now follows from Pythagoras' theorem applied to T.)

An interesting feature of this statement is that it makes no mention of the fact that the points were supposed to be three-dimensional.

We have therefore stumbled on a method for calculating distances in *any* number of dimensions. For example, the distance between the two points (1, 0, –1, 4, 2) and (3, 1, 1, 1, –1) (which lie in five-dimensional space) is

$$\sqrt{(3-1)^2 + (1-0)^2 + (1-(-1))^2 + (1-4)^2 + (-1-2)^2} =$$
$$\sqrt{4+1+4+9+9} = \sqrt{27}$$

Now this way of putting things is a little misleading because it suggests that there was always a distance between any pair of five-dimensional points (remember that a five-dimensional point means nothing more than a sequence of five real numbers) and that we have discovered how to work these distances out. Actually, however, what we have done is *define* a notion of distance. No physical reality forces us to decide that five-dimensional distance should be calculated in the manner described. On the other hand, this method is so clearly the natural generalization of what we do in two and three dimensions that it would be strange to adopt any other definition.

Once distance has been defined, we can begin to generalize other concepts. For example, a sphere is clearly the three-dimensional equivalent of a circle. What would a four-dimensional 'sphere' be? As with distance, we can answer this question if we can describe the two- and three-dimensional versions in a way that does not actually mention the number of dimensions. This is not at all hard: circles and spheres can both be described as the set of all points at a fixed distance (the radius) from some given point (the centre). There is nothing to stop us from using exactly the same definition for four-dimensional spheres, or eighty-seven-dimensional spheres for that matter. For example, the four-dimensional sphere of radius 3 about the point (1, 1, 0, 0) is the set of all (four-dimensional) points at distance 3 from (1, 1, 0, 0). A four-dimensional point is a sequence (a, b, c, d) of four real numbers. Its distance from (1, 1, 0, 0) is (according to our earlier definition)

$$\sqrt{(a-1)^2 + (b-1)^2 + c^2 + d^2}$$

Therefore, another description of this four-dimensional sphere is that it is the set of all quadruples (a, b, c, d) for which

$$\sqrt{(a-1)^2 + (b-1)^2 + c^2 + d^2} = 3$$

For example, $(1, -1, 2, 1)$ is such a quadruple, so this is a point in the given four-dimensional sphere.

Another concept that can be generalized is that of a square in two dimensions and a cube in three. As Figure 23 makes clear, the set of all points (a, b) such that both a and b lie between 0 and 1 forms a square of side length 1, and its four vertices are the points $(0, 0)$, $(0, 1)$, $(1, 0)$, and $(1, 1)$. In three dimensions one can define a cube by taking all points (a, b, c) such that a, b, and c all lie between 0 and 1. Now there are eight vertices: $(0, 0, 0)$, $(0, 0, 1)$, $(0, 1, 0)$, $(0, 1, 1)$, $(1, 0, 0)$, $(1, 0, 1)$, $(1, 1, 0)$, and $(1, 1, 1)$. Similar definitions are obviously possible in higher dimensions. For example, one can obtain a six-dimensional cube, or rather a mathematical construction clearly worthy of the name, by taking all points (a, b, c, d, e, f) with all their coordinates lying between 0 and 1. The vertices will be all points for which every coordinate is 0 or 1: it is not hard to see that the number of vertices doubles each time you add a dimension, so in this case there are 64 of them.

One can do much more than simply *define* shapes. Let me illustrate this briefly by calculating the number of edges of a five-dimensional cube. It is not immediately obvious what an edge is, but for this we can take our cue from what happens in two and three dimensions: an edge is the line that joins two neighbouring vertices, and two vertices are regarded as neighbours if they differ in precisely one coordinate. A typical vertex in the five-dimensional cube is a point such as $(0, 0, 1, 0, 1)$ and, according to the definition just given, its neighbours are $(1, 0, 1, 0, 1)$, $(0, 1, 1, 0, 1)$, $(0, 0, 0, 0, 1)$, $(0, 0, 1, 1, 1)$, and $(0, 0, 1, 0, 0)$. In general, each vertex has five neighbours, and hence five edges coming out of it. (I leave it to the reader to generalize from two and three dimensions the notion of the line joining two neighbouring vertices. For this calculation it does not matter.) Since there are $2^5 = 32$ vertices, it looks as though there are $32 \times 5 = 160$ edges. However, we have counted each edge twice – once each for its two end points – so the correct answer is half of 160, that is, 80.

One way of summarizing what we are doing is to say that we are converting geometry into algebra, using coordinates to translate geometrical concepts into equivalent concepts that involve only relationships between numbers. Although we cannot directly generalize the geometry, we *can* generalize the algebra, and it seems reasonable to call this generalization higher-dimensional geometry. Obviously, five-dimensional geometry is not as directly related to our immediate experience as three-dimensional geometry, but this does not make it impossible to think about, or prevent it from being useful as a model.

Can four-dimensional space be visualized?

In fact, the seemingly obvious statement that it is possible to visualize three-dimensional objects but not four-dimensional ones does not really stand up to close scrutiny. Although visualizing an object feels rather like looking at it, there are important differences between the two experiences. For example, if I am asked to visualize a room with which I am familiar, but not very familiar, I have no difficulty in doing so. If I am then asked simple questions about it, such as how many chairs it contains or what colour the floor is, I am often unable to answer them. This shows that, whatever a mental image is, it is not a photographic representation.

In a mathematical context, the important difference between being able to visualize something and not being able to is that in the former case one can somehow answer questions directly rather than having to stop and calculate. This directness is of course a matter of degree, but it is no less real for that. For example, if I am asked to give the number of edges of a three-dimensional cube, I can do it by 'just seeing' that there are four edges round the top, four round the bottom, and four going from the top to the bottom, making twelve in all.

In higher dimensions, 'just seeing' becomes more difficult, and one is often forced to argue more as I did when discussing the analogous question in five dimensions. However, it is sometimes possible. For example, I can think of a four-dimensional cube as consisting of two three-dimensional cubes facing each other, with corresponding vertices joined by edges (in the fourth dimension), just as a three-dimensional cube consists of two squares facing each other with corresponding vertices joined. Although I do not have a completely clear picture of four-dimensional space, I can still 'see' that there are twelve edges for each of the two three-dimensional cubes, and eight edges linking their vertices together. This gives a total of 12 + 12 + 8 = 32. Then I can 'just see' that a five-dimensional cube is made up of two of these, again with corresponding vertices linked, making a total of 32 + 32 + 16 = 80 edges (32 for each four-dimensional cube and 16 for the edges between them), exactly the answer I obtained before. Thus, I have some rudimentary ability to visualize in four and five dimensions. (If you are bothered by the word 'visualize' then you can use another one, such as 'conceptualize'.) Of course, it is much harder than visualizing in three dimensions – for example, I cannot directly answer questions about what happens when you rotate a four-dimensional cube, whereas I can for a three-dimensional one – but it is also distinctly easier than fifty-three-dimensional visualization, which it could not be if they were both impossible. Some mathematicians specialize in four-dimensional geometry, and their powers of four-dimensional visualization are highly developed.

This psychological point has an importance to mathematics that goes well beyond geometry. One of the pleasures of devoting one's

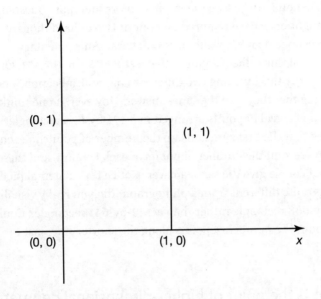

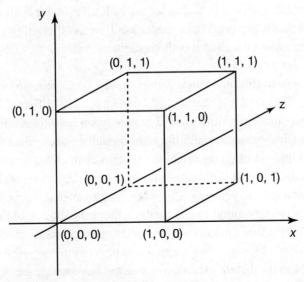

23. A unit square and a unit cube

life to mathematical research is that, as one gains in expertise, one finds that one can 'just see' answers to more and more questions that might once have required an hour or two of hard thought, and the questions do not have to be geometrical. An elementary example of this is the statement that $471 \times 638 = 638 \times 471$. One could verify this by doing two different long multiplications and checking that they give the same answer. However, if one thinks instead of a grid of points arranged in a 471-by-638 rectangle, one can see that the first sum adds up the number of points in each row and the second the number of points in each column, and these must of course give the same answer. Notice that a mental picture here is quite different from a photograph: did you really visualize a 471-by-638 rectangle rather than a 463-by-641 rectangle? Could you count the number of points along the shorter side, just to check?

What is the point of higher-dimensional geometry?

It is one thing to demonstrate that some sense can be made of the idea of higher-dimensional geometry, but quite another to explain why it is a subject worth taking seriously. Earlier in the chapter, I claimed that it was useful as a model, but how can this be, given that the actual space we inhabit is three-dimensional?

The answer to this question is rather simple. One of the points I made in Chapter 1 was that a model can have many different uses. Even two- and three-dimensional geometry are used for many purposes other than a straightforward modelling of physical space. For example, we often represent the motion of an object by drawing a graph that records the distance it has travelled at different times. This graph will be a curve in the plane, and geometrical properties of the curve correspond to information about the motion of the object. Why is two-dimensional geometry appropriate for modelling this motion? Because there are two numbers of interest – the time elapsed and the distance travelled – and, as I have said, one can think of two-dimensional space as the collection of all pairs of numbers.

This gives us a clue about why higher-dimensional geometry can be useful. There may not be any high-dimensional space lurking in the universe, but there are plenty of situations in which we need to consider collections of several numbers. I shall very briefly describe two, after which it should become obvious that there are many more.

Suppose that I wish to describe the position of a chair. If it is standing upright, then its position is completely determined by the points where two of its legs meet the floor. These two points can each be described by two coordinates. The result is that four numbers can be used to describe the position of the chair. However, these four numbers are related, because the distance between the bottoms of the legs is fixed. If this distance is d and the legs meet the floor at the points (p, q) and (r, s) then $(p - r)^2 + (q - s)^2 = d^2$, by Pythagoras' theorem. This puts a constraint on p, q, r, and s, and one way to describe this constraint uses geometrical language: the point (p, q, r, s), which belongs to four-dimensional space, is forced to lie in a certain three-dimensional 'surface'. More complicated physical systems can be analysed in a similar way, and the dimensions become much higher.

Multi-dimensional geometry is also very important in economics. If, for example, you are wondering whether it is wise to buy shares in a company, then much of the information that will help you make your decision comes in the form of numbers – the size of the workforce, the values of various assets, the costs of raw materials, the rate of interest, and so on. These numbers, taken as a sequence, can be thought of as a point in some high-dimensional space. What you would like to do, probably by analysing many other similar companies, is identify a region of that space, the region where it is a good idea to buy the shares.

Fractional dimension

If one thing seems obvious from the discussion so far, it is that the dimension of any shape will always be an integer. What could it possibly mean to say that you need two and a half coordinates to specify a point – even a mathematical one?

This argument may look compelling, but we were faced with a very similar difficulty before we defined the number $2^{3/2}$ in Chapter 2, and managed to circumvent it by using the abstract method. Can we do something similar for dimension? If we want to, then we must find some property associated with dimension that does not immediately imply that it is a whole number. This rules out anything to do with the number of coordinates, which seems so closely bound up with the very idea of dimension that it is hard to think of anything else. There is, however, another property, mentioned briefly at the beginning of this chapter, that gives us exactly what we need.

An important aspect of geometry which varies with dimension is the rule that determines what happens to the size of a shape when you expand it by a factor of t in every direction. By size, I mean length, area, or volume. In one dimension, the size multiplies by t, or t^1, in two dimensions it multiplies by t^2, and in three it multiplies by t^3. Thus, the power of t tells us the dimension of the shape.

So far we have not quite managed to banish whole numbers from the picture, because the numbers two and three are implicit in the words 'area' and 'volume'. However, we can do without these words as follows. Why is it that a square of side length three has nine times the area of a square of side length one? The reason is that one can divide the larger square into nine congruent copies of the smaller one (see Figure 24.). Similarly, a three-by-three-by-three cube can be divided into twenty-seven one-by-one-by-one cubes, so its volume is twenty-seven times that of a one-by-one-by-one cube. So we can say that a cube is three-dimensional because if it is expanded

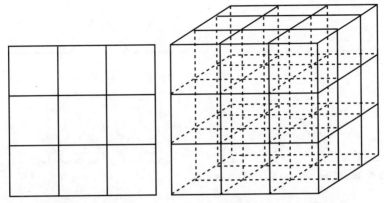

24. Dividing a square into $9 = 3^2$ smaller squares, and a cube into $27 = 3^3$ smaller cubes

by a factor of t, where t is a whole number greater than 1, then the new cube can be divided into t^3 copies of the old one. Note that the word 'volume' did not appear in the last sentence.

We may now ask: is there a shape for which we can reason as above and obtain an answer that is not an integer? The answer is yes. One of the simplest examples is known as the Koch snowflake. It is not possible to describe it directly: instead, it is defined as the limit of the following process. Start with a straight line segment, of length one, say. Then divide it into three equal pieces, and replace the middle piece with the other two sides of the equilateral triangle that has this middle piece as its base. The result is a shape built out of four straight line segments, each of length one-third. Divide each of these into three equal pieces, and again replace each middle piece by the other two sides of an equilateral triangle. Now the result is a shape made of sixteen line segments, each of length one-ninth. It is clear how to continue this process: the first few stages are illustrated in Figure 25. It is not too hard to prove rigorously that this process leads to a limiting shape, as the pictures suggest, and this shape is the Koch snowflake. (It looks more like a snowflake if you take three copies of it and put them together round a triangle.)

25. Building up the Koch snowflake

The Koch snowflake has several interesting features. The one that will concern us is that it can be built out of smaller copies of itself. Once again, this can be seen from the picture: it consists of four copies, and each copy is a shrinking of the whole shape by a factor of one-third. Let us now consider what that tells us about its dimension.

If a shape is d-dimensional, then when we shrink it by a factor of one-third its size should go down by a factor of 3^d. (As we have seen, this is true when d is 1, 2, or 3.) Thus, if we can build it out of smaller copies, then we should need 3^d of them. Since four copies are needed for the Koch snowflake, its dimension d should be the number for which $3^d = 4$. Since $3^1 = 3$ and $3^2 = 9$, this means that d lies between 1 and 2, and so is not a whole number. In fact, it is $\log_3 4$, which is approximately 1.2618595.

This calculation depends on the fact that the Koch snowflake can be decomposed into smaller copies of itself, which is a very unusual feature: even a circle doesn't have it. However, it is possible to develop the above idea and give a definition of dimension that is much more widely applicable. As with our other uses of the abstract method, this does not mean that we have discovered the 'true dimension' of the Koch snowflake and similar exotic shapes – but merely that we have found the only possible definition consistent with certain properties. In fact, there are other ways of defining dimension that give different answers. For example, the Koch snowflake has a 'topological dimension' of 1. Roughly speaking, this is because, like a line, it can be broken into two disconnected parts by the removal of any one of its interior points.

This sheds interesting light on the twin processes of abstraction and generalization. I have suggested that to generalize a concept one should find some properties associated with it and generalize those. Often there is only one natural way to do this, but sometimes different sets of properties lead to different generalizations, and sometimes more than one generalization is fruitful.

Chapter 6
Geometry

Perhaps the most influential mathematics book of all time is Euclid's *Elements*, written in around 300 BC. Although Euclid lived over two thousand years ago, he was in many ways the first recognizably modern mathematician – or at least the first that we know about. In particular, he was the first author to make systematic use of the axiomatic method, beginning his book with five axioms and deducing from them a large body of geometrical theorems. The geometry with which most people are familiar, if they are familiar with any at all, is the geometry of Euclid, but at the research level the word 'geometry' has a much broader definition: today's geometers do not spend much of their time with a ruler and compass.

Euclidean geometry

Here are Euclid's axioms. I follow the normal convention and use the word 'line' for a line that extends indefinitely in both directions. A 'line segment' will mean a line with two end-points.

1. Any two points can be joined by exactly one line segment.
2. Any line segment can be extended to exactly one line.
3. Given any point P and any length r, there is a circle of radius r with P as its centre.
4. Any two right angles are congruent.

5. If a straight line N intersects two straight lines L and M, and if the interior angles on one side of N add up to less than two right angles, then the lines L and M intersect on that side of N.

The fourth and fifth axioms are illustrated in Figure 26. The fourth means that you can slide any right angle until it lies exactly on any other. As for the fifth, because the angles marked a and β add up to less than 180 degrees, it tells us that the lines L and M meet somewhere to the right of N. The fifth axiom is equivalent to the so-called 'parallel postulate', which asserts that, given any line L and any point x not lying on L, there is exactly one line M that goes through x and never meets L.

Euclid used these five axioms to build up the whole of geometry as it was then understood. Here, for example, is an outline of how to prove the well-known result that the angles of a triangle add up to 180 degrees. The first step is to show that if a line N meets two parallel lines L and M, then opposite angles are equal. That is, in something like Figure 27 one must have $a = a'$ and $\beta = \beta'$. This is a consequence of the fifth axiom. First, it tells us that $a' + \beta$ is at least 180, or otherwise L and M would have to meet (somewhere to the left of line N in the diagram). Since a and β together make a straight line, $\beta = 180 - a$, so it follows that $a' + (180 - a)$ is at least 180, which means that a' is at least as big as a. By the same argument, $a + \beta' = a + (180 - a')$ must be at least 180, so a is at least as big as a'. The only way this can happen is if a and a' are equal. Since $\beta = 180 - a$ and $\beta' = 180 - a'$, it follows that $\beta = \beta'$ as well.

Now let ABC be a triangle, and let the angles at A, B, and C be a, β, and γ respectively. By the second axiom we can extend the line segment AC to a line L. The parallel postulate tells us that there is a line M through B that does not meet L. Let a' and γ' be the angles marked in Figure 28. By what we have just proved, $a' = a$ and $\gamma' = \gamma$. It is clear that $a' + \beta + \gamma' = 180$, since the three angles a', β, and γ' together make a straight line. Therefore $a + \beta + \gamma = 180$, as required.

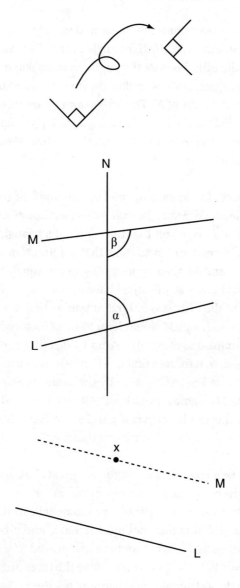

26. Euclid's fourth axiom and two versions of the fifth

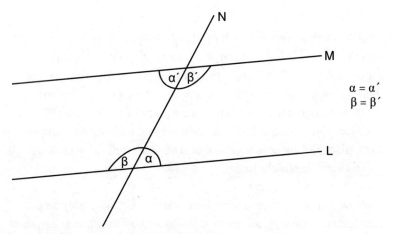

27. A consequence of Euclid's fifth axiom

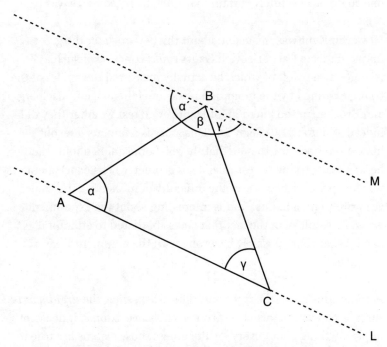

28. A proof that the angles of a triangle add up to 180 degrees

What does this argument tell us about everyday life? An obvious conclusion seems to be that if you take three points *A*, *B*, and *C* in space, and carefully measure the three angles of the triangle *ABC*, then they will add up to 180 degrees. A simple experiment will confirm this: just draw a triangle on a piece of paper, cut it out as neatly as you can, tear it into three pieces containing one corner each, place the corners together, and observe that the angles do indeed form a straight line.

If you are now convinced that there could not conceivably be a physical triangle with angles that fail to add up to 180 degrees, then you are in good company, as this was the conclusion drawn by everybody from Euclid in 300 BC to Immanuel Kant at the end of the 18th century. Indeed, so convinced was Kant that he devoted a significant part of his *Critique of Pure Reason* to the question of how one could be absolutely certain that Euclidean geometry was true.

However, Kant was mistaken: about thirty years later the great mathematician Carl Friedrich Gauss *could* conceive of such a triangle, as a result of which he actually measured the angles of the triangle formed by the mountain peaks of Hohenhagen, Inselberg, and Brocken in the kingdom of Hanover, to test whether they did indeed add up to 180 degrees. (This story is a famous one, but honesty compels me to note that there is some doubt about whether he was really trying to test Euclidean geometry.) His experiment was inconclusive because of the difficulty of measuring the angles accurately enough, but what is interesting about the experiment is less its result than the fact that Gauss bothered to attempt it at all. What could possibly be wrong with the argument I have just given?

Actually, this is not the right question to ask, since the *argument* is correct. However, since it rests on Euclid's five axioms, it does not imply anything about everyday life unless those axioms are true in everyday life. Therefore, by questioning the truth of Euclid's axioms one can question the *premises* of the argument.

But which of the axioms looks in the least bit dubious? It is difficult to find fault with any of them. If you want to join two points in the real world by a line segment, then all you have to do is hold a taut piece of string in such a way that it goes through both points. If you want to extend that line segment to a straight line, then you could use a laser beam instead. Similarly, there seems to be no difficulty with producing circles of any desired radius and centre, and experience shows that if you take two right-angled corners of paper, you can place one of them exactly over the other one. Finally, what is to stop two lines going off into the distance for ever, like a pair of infinitely long railway tracks?

The parallel postulate

Historically, the axiom that caused the most suspicion, or at least uneasiness, was the parallel postulate. It is more complicated than the other axioms, and involves the infinite in a fundamental way. Is it not curious, when one proves that the angles of a triangle add up to 180 degrees, that the proof should depend on what happens in the outermost reaches of space?

Let us examine the parallel postulate more carefully, and try to understand why it feels so obviously true. Perhaps in the back of our minds is one of the following arguments.

(1) Given a straight line L and a point x not on it, all you have to do to produce a parallel line through x is choose the line through x that goes *in the same direction as L*.
(2) Let y be another point, on the same side of L as x and at the same distance from L. Join x to y by a line segment (axiom 1) and then extend this line segment to a full line M (axiom 2). Then M will not meet L.
(3) Let M be the straight line consisting of all points on the same side of L as x and at the same distance. Obviously this does not meet L.

The arguments so far concern the *existence* of a line parallel to L. Here is a more complicated argument aimed at showing that there can be at most one such line, which is the other part of the parallel postulate.

(4) Join L to M by evenly spaced perpendicular line segments (making the railway tracks of Figure 29) with one of these segments through x. Now suppose that N is another line through x. On one side of x, the line N must lie between L and M, so it meets the next line segment at a point u, which is between L and M. Suppose for the sake of example that u is 1% of the way along the line segment from M to L. Then N will meet the next line segment 2% of the way along, and so on. Thus, after 100 segments, N will have met L. Since all we have assumed about N is that it is not M, it follows that M is the only line through x that does not meet L.

Finally, here is an argument that appears to show both the existence and the uniqueness of a line parallel to L through a given point.

(5) A point in the plane can be described by Cartesian coordinates. A (non-vertical) line L has an equation of the form $y = mx + c$. By

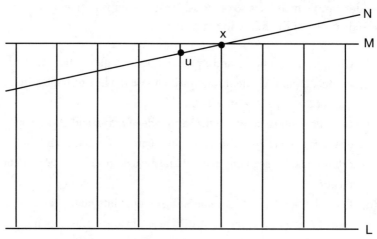

29. The uniqueness of parallel lines

varying c we can move L up and down. Obviously, no two of the lines thus produced can cross each other, and every point is contained in exactly one of them.

Notice that what I have just done is try to *prove* the parallel postulate, and that is exactly what many mathematicians before the 19th century tried to do. What they most wanted was to deduce it from the other four axioms, thus showing that one could dispense with it. However, nobody managed to do so. The trouble with the arguments I have just given, and others like them, is that they contain hidden assumptions which, when one makes them explicit, are not obvious consequences of Euclid's first four axioms. Though plausible, they are no *more* plausible than the parallel postulate itself.

Spherical geometry

A good way to bring these hidden assumptions out into the open is to examine the same arguments applied in a different context, one in which the parallel postulate is definitely not true. With this in mind, let us think for a moment about the surface of a sphere.

It is not immediately obvious what it means to say that the parallel postulate is untrue on the surface of a sphere, because the surface of a sphere contains no straight lines at all. We shall get round this difficulty by applying an idea of fundamental importance in mathematics. The idea, which is a profound example of the abstract method at work, is to *reinterpret* what is meant by a straight line, so that the surface of a sphere does contain straight lines after all.

There is in fact a natural definition: a line segment from x to y is the shortest path from x to y *that lies entirely within the surface of the sphere*. One can imagine x and y as cities and the line segment as the shortest route that an aeroplane could take. Such a path will be part of a 'great circle', which means a circle obtained by taking a plane through the centre of the sphere and seeing where it cuts the surface (see Figure 30). An example of a great circle is the equator of the earth (which, for the purposes of discussion, I shall take to be a perfect sphere). Given the way that we have defined line segments, a great circle makes a good definition of a 'straight line'.

If we adopt this definition, then the parallel postulate is certainly false. For example, let L be the earth's equator and let x be a point in the northern hemisphere. It is not hard to see that any great circle through x will lie half in the northern hemisphere and half in the southern, crossing the equator at two points that are exactly opposite each other (see Figure 31). In other words, there is no line (by which I still mean great circle) through x that does not meet L.

This may seem a cheap trick: if I define 'straight line' in a new way, then it is not particularly surprising if the parallel postulate ceases to hold. But it is not meant to be surprising – indeed the definition was designed for that very purpose. It becomes interesting when we examine some of the attempted *proofs* of the parallel postulate. In each case, we will discover an assumption that is not valid for spherical geometry.

For example, argument (1) assumes that it is obvious what is meant by the phrase 'the same direction'. But on the surface of a sphere this is not obvious at all. To see this, consider the three points P, Q, and N shown in Figure 32. N is the North Pole, P lies on the equator, and Q also lies on the equator, a quarter of the way round from P. Also on Figure 32 is a small arrow at P, pointing along the equator towards Q. Now what arrow could one draw at Q going in the same direction? The natural direction to choose is still along the equator, away from P. What about an arrow at N in the same direction again? We could choose this as follows. Draw the line segment from P to N. Since the arrow at P is at right angles to this line segment, the arrow at N should be as well, which means, in fact, that it points down towards Q. However, we now have a problem, which is that

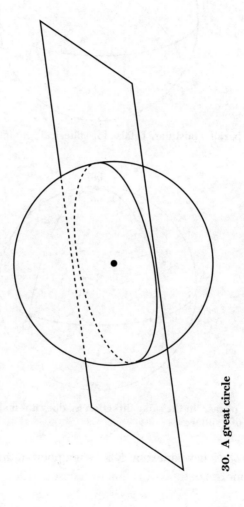

30. A great circle

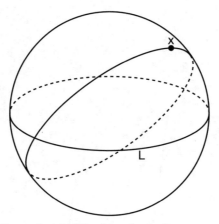

31. The parallel postulate is false for spherical geometry

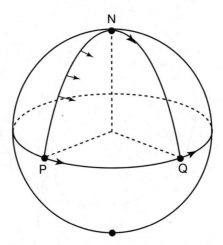

32. The phrase 'in the same direction as' does not make sense on the surface of a sphere

the arrow we have drawn at N does not point in the same direction as the one at Q.

The trouble with argument (2) is that it is not detailed enough. *Why* will the line M defined there not meet L? After all, if L and M are spherical lines then they *will* meet. As for argument (3), it assumes that M is a straight line. This is untrue for the sphere: if L is the equator and M consists of all points that are 1,000 miles north of the equator, then M is *not* a great circle. In fact, it is a line of constant latitude, which, as any pilot or sailor will tell you, does not give the shortest route between two points.

Argument (4) is a little different, since it is concerned with the uniqueness of parallel lines rather than their existence. I shall discuss this in the next section. Argument (5) makes a huge assumption: that space can be described by Cartesian coordinates. Again, this is untrue of the surface of the sphere.

The point about introducing the sphere is that it enables us to isolate from each of the arguments (1), (2), (3) and (5) some assumption that effectively says, 'The geometry we are doing is not spherical geometry.' You might wonder what is wrong with this: after all, we are *not* doing spherical geometry. You might also wonder how one could ever hope to show that the parallel postulate does not follow from the rest of Euclid's axioms, if indeed it doesn't. It is no good saying that mathematicians have tried to deduce it for centuries without success. How can we be sure that some young genius in two hundred years' time will not have a wonderful new idea which finally leads to a proof?

This question has a beautiful answer, at least in principle. Euclid's first four axioms were devised to describe the geometry of an infinite, flat, two-dimensional space, but we are not obliged to interpret them that way, unless, of course, this flatness actually follows from the axioms. If we could somehow reinterpret (one might almost say 'misinterpret') the axioms by giving new meanings to phrases such as 'line segment', rather as we have done with spherical geometry, and if when we had done so we found that the first four axioms were true but the parallel postulate was false, then we would have shown that the parallel postulate does not follow from the other axioms.

To see why this is so, imagine a purported proof, which starts with Euclid's first four axioms and arrives, after a series of strict logical steps, at the parallel postulate. Since the steps follow as a matter of logic, they will remain valid even when we have given them their new interpretation. Since the first four axioms are true under the new interpretation and the parallel postulate is not true, there must be a mistake in the argument.

Why do we not just use spherical geometry as our reinterpretation? The reason is that, unfortunately, not all of the first four of Euclid's axioms are true in the sphere. For example, a sphere does not contain circles of arbitrarily large radius, so axiom 3 fails, and there is not just one shortest route from the North to the South Pole, so axiom 1 is not true either. Hence, although spherical geometry helped us understand the defects of certain attempted proofs of the parallel postulate, it still leaves open the possibility that some other proof might work. Therefore, I shall turn to another reinterpretation, called hyperbolic geometry. The parallel postulate will again be false, but this time axioms 1 to 4 will all be true.

Hyperbolic geometry

There are several equivalent ways of describing hyperbolic geometry; the one I have chosen is known as the disc model, which was discovered by the great French mathematician Henri Poincaré. While I cannot define it precisely in a book like this, I can at least explain some of its main features and discuss what it tells us about the parallel postulate.

Understanding the disc model is more complicated than understanding spherical geometry because one has to reinterpret not only the terms 'line' and 'line segment', but also the idea of distance. On the surface of the sphere, distance has an easily grasped definition: the distance between two points x and y is the shortest possible length of a path from x to y that lies within the surface of the sphere. Although a similar definition holds for hyperbolic geometry, it is not obvious, for reasons that will become clear, what the shortest path is, or indeed what the length of *any* path is.

Figure 33 shows a tessellation of the hyperbolic disc by regular pentagons. Of course, this statement has to be explained, since it is untrue if we understand distance in the usual way: the edges of these 'pentagons' are visibly not straight and do not have the same length. However, distances in the hyperbolic disc are not defined in the usual way, and become larger, relative to normal distance, as you approach the boundary. Indeed, they become so much larger that the boundary is, despite appearances, infinitely far from the centre. Thus, the reason that the pentagon marked with an asterisk appears to have one side longer than all the others is that that side is closer to the centre. The other sides may look shorter, but hyperbolic distance is defined in such a way that this apparent shortness is exactly compensated for by their being closer to the edge.

33. A tessellation of the hyperbolic plane by regular pentagons

If this seems confusing and paradoxical, then think of a typical map of the world. As everyone knows, because the world is round and the map is flat, distances are necessarily distorted. There are various ways of carrying out the distortion, and in the most common one, Mercator's projection, countries near the poles appear to be much larger than they really are. Greenland, for example, seems to be comparable in size to the whole of South America. The nearer you are to the top or bottom of such a map, the smaller the distances are, compared with what they appear to be.

A well-known effect of this distortion is that the shortest route between two points on the earth's surface appears, on the map, to be curved. This phenomenon can be understood in two ways. The first is to forget the map and visualize a globe instead, and notice that if you have two points in the northern hemisphere, the first a long way east of the other (a good example is Paris and Vancouver), then the shortest route from the first to the second will pass close to the North Pole rather than going due west. The second is to argue from the original map and reason that if distances near the top of the map are shorter than they appear, then one can shorten the journey by going somewhat north as well as west. It is difficult to see in this way precisely what the shortest path will be, but at least the principle is clear that a 'straight line' (from the point of view of spherical distances) will be curved (from the point of view of the distances on the actual map).

As I have said, when you approach the edge of the hyperbolic disc, distances become *larger* compared with how they look. As a result of this, the shortest path between two points has a tendency to deviate towards the centre of the disc. This means that it is not a straight line in the usual sense (unless that line happens to pass exactly through the centre). It turns out that a hyperbolic straight line, that is, a shortest path from the point of view of hyperbolic geometry, is the arc of a circle that meets the boundary of the main circle at right angles (see Figure 34). If you now look again at the pentagonal tessellation of Figure 35, you will see that the edges of

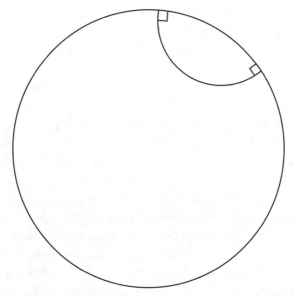

34. A typical hyperbolic line

the pentagons, though they do not appear straight, are in fact hyperbolic line segments since they can be extended to hyperbolic straight lines, according to the definition I have just given. Similarly, although the pentagons do not seem to be all of the same size and shape, they are, since the ones near the edge are far bigger than they seem – the opposite of what happens with Greenland. Thus, just like Mercator's projection, the disc model is a distorting 'map' of actual hyperbolic geometry.

It is natural to ask at this point what actual hyperbolic geometry is like. That is, what is the distorting map a map *of*? What stands in relation to the disc model as the sphere does to Mercator's projection? The answer to this is rather subtle. In a way it is a fluke that spherical geometry can be realized as a surface that sits in three-dimensional space. If we had *started* with Mercator's projection, with its strange notion of distances, without knowing that what we had was a map of the sphere, then we would have been surprised and delighted to discover that there happened to be a beautifully symmetrical surface in space, a map of this map, so to speak, where distances were particularly simple, being nothing but the lengths of shortest paths in the usual, easily understood sense.

Unfortunately, nothing like this exists for hyperbolic geometry. Yet, curiously, this does not make hyperbolic geometry any less real than spherical geometry. It makes it harder to understand, at least initially, but as I stressed in Chapter 2 the reality of a mathematical concept has more to do with what it does than with what it is. Since it is possible to say what the hyperbolic disc does (for example, if you asked me what it would mean to rotate the pentagonal tessellation through 30 degrees about one of the vertices of the central pentagon, then I could tell you), hyperbolic geometry is as real as any other mathematical concept. Spherical geometry may be easier to understand from the point of view of three-dimensional Euclidean geometry, but this is not a fundamental difference.

Another of the properties of hyperbolic geometry is that it satisfies the first four of Euclid's axioms. For example, any two points can be joined by exactly one hyperbolic straight line segment (that is, arc of a circle that cuts the main circle at right angles). It may seem as though you cannot find a circle of large radius about any given point, but to think that is to have forgotten that distances become larger near the edge of the disc. In fact, if a hyperbolic circle almost brushes the edge, then its radius (its hyperbolic radius, that is) will be very large indeed. (Hyperbolic circles happen to look like ordinary circles, but their centres are not where one expects them to be. See Figure 35.)

As for the parallel postulate, it is false for hyperbolic geometry, just as we hoped. This can be seen in Figure 36, where I have marked three of the (hyperbolic) lines L, M_1, and M_2. The lines M_1 and M_2 meet at a point marked x, but neither of them meets L. Thus, there are two lines through x (and in fact infinitely many) that do not meet L. This contradicts the parallel postulate, which stipulates that there should be only one. In other words, in hyperbolic geometry we

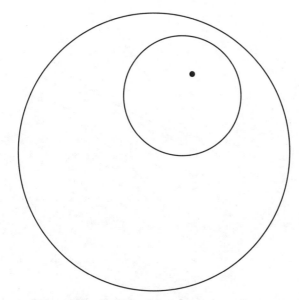

35. A typical hyperbolic circle, and its centre

have exactly the alternative interpretation of the Euclidean axioms that we were looking for in order to show that the parallel postulate was not a consequence of the other four axioms.

Of course, I have not actually proved in this book that hyperbolic geometry has all the properties I have claimed for it. To do so takes a few lectures in a typical university mathematics course, but I can at least say more precisely how to define hyperbolic distance. To do this, I must specify *by how much* distances near the edge of the disc are larger than they appear. The answer is that hyperbolic distances at a point P are larger than 'normal' distances by $1/d^2$, where d is the distance from P to the boundary of the circle. To put that another way, if you were to move about in the hyperbolic disc, then your speed as you passed P would, according to hyperbolic notions of distance, be $1/d^2$ times your apparent speed, which means that if you maintained a constant hyperbolic speed, you would appear to move more and more slowly as you approached the boundary of the disc.

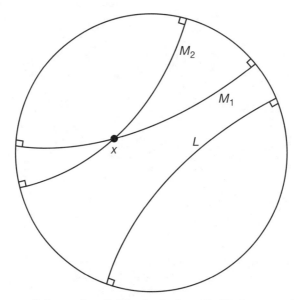

36. The parallel postulate is false in the hyperbolic plane

Just before we leave hyperbolic geometry, let us see why argument (4) that I gave earlier fails to prove the uniqueness of parallel lines. The idea was that, given a line L, a point x not on L, and a line M through x that did not meet L, one could join L to M by several line segments that were perpendicular to both L and M, dividing up the space between L and M into rectangles. It seems obvious that one can do this, but in the hyperbolic world it is not possible because the angles of a quadrilateral always add up to *less* than 360 degrees. In other words, in the hyperbolic disc the rectangles needed for the argument simply do not exist.

How can space be curved?

One of the most paradoxical-sounding phrases in mathematics (and physics) is 'curved space'. We all know what it means for a line or surface to be curved, but space itself just *is*. Even if one could somehow make sense of the idea of three-dimensional curviness, the analogy with a curved surface suggests that we would not be able to see for ourselves whether space was curved unless we could step out into a fourth dimension to do so. Perhaps we would then discover that the universe was the three-dimensional surface of a four-dimensional sphere (a notion I explained in Chapter 5), which at least *sounds* curved.

Of course, all this is impossible. Since we do not know how to stand outside the universe – the very idea is almost a contradiction in terms – the only evidence we can use comes from within it. So what evidence might persuade us that space is curved?

Yet again, the question becomes easier if one takes an abstract approach. Instead of engaging in extraordinary mental gymnastics as we try to apprehend the true nature of curvy space, let us simply follow the usual procedure for generalizing mathematical concepts. We understand the word 'curved' when it applies to two-dimensional surfaces. In order to use it in an unfamiliar context, that is, a three-dimensional one, we must try to find *properties* of curved surfaces that will generalize easily, just as we did in order to define $2^{3/2}$, or five-dimensional cubes, or the dimension of the Koch snowflake. Since the sort of property we wish to end up with is one that can be detected from *within* space, we ought to look at ways of detecting the curvature of a surface that do not depend on standing outside it.

How, for example, can we convince ourselves that the surface of the earth is curved? One way is to go up in a space shuttle, look back, and see that it is approximately spherical. However, the following experiment, which is much more two-dimensional, would also be very persuasive. Start at the North Pole and travel due south for about 6,200 miles, having marked your initial direction. Then turn to your left and go the same distance again. Then turn to your left and go the same distance one more time. 6,200 miles is roughly the distance from the North Pole to the equator, so your journey will have taken you from the North Pole to the equator, a quarter of the way round the equator, and back to the North Pole again. Moreover, the direction at which you arrive back will be at right angles to your starting direction. It follows that on the earth's surface there is an equilateral triangle with all its angles equal to a right angle. On a flat surface, the angles of an equilateral triangle have to be 60 degrees, as they are all equal and add up to 180, so the surface of the earth is not flat.

Thus, one way of demonstrating that a two-dimensional surface is curved, from within that surface, is to find a triangle whose angles do not add up to 180 degrees, and this is something that can be attempted in three dimensions as well. I have concentrated in this chapter on Euclidean, spherical, and hyperbolic geometry in two dimensions, but they can be generalized quite easily to three dimensions. If we measure the angles of triangles in space and find that they add up to more than 180 degrees, then that will suggest that space is more like a three-dimensional version of the surface of a sphere than like the sort of space that can be described by three Cartesian coordinates.

If this happens, then it seems reasonable to say that space is positively curved. Another feature that one would expect of such a space is that lines that started off in the same direction would converge and eventually meet. Still another is that the circumference of a circle of radius r would not be $2\pi r$, but a little less.

You may be tempted to point out that space as we know it does not have these peculiarities. Lines that begin in the same direction continue in the same direction, and the angles of triangles and circumferences of circles are what they ought to be. In other words, it appears that, even though it is logically possible for space to be curved, as a matter of fact it is flat. However, it could be that space appears to be flat to us only because we inhabit such a small part of it, just as the earth's surface appears to be flat, or rather flat with bumps of various sizes, to somebody who has not travelled far.

In other words, it may be that space is only *roughly* flat. Perhaps if we could form a very large triangle then we would find that its angles did not add up to 180 degrees. This, of course, was what Gauss attempted, but his triangle was nowhere near large enough. However, in 1919, one of the most famous scientific experiments of all time showed that the idea of curved space was not just a fantasy of mathematicians, but a fact of life. According to Einstein's general theory of relativity, which was published four years earlier, space is curved by gravity, and therefore light does not always travel in a straight line, at least as Euclid would understand the term. The effect is too small to be detected easily, but the opportunity came in 1919 with a total eclipse of the sun, visible from Principe Island in the Gulf of Guinea. While it was happening, the physicist Arthur Eddington took a photograph that showed the stars just next to the sun in not quite their expected places, exactly as Einstein's theory had predicted.

Though it is now accepted that space (or, more accurately, spacetime) is curved, it could be that, like the mountains and valleys on the earth's surface, the curvature that we observe is just a small perturbation of a much larger and more symmetrical shape. One of the great open questions of astronomy is to determine the *large-scale* shape of the universe, the shape that it would have if one ironed out the curves due to stars, black holes, and so on. Would it still be curved, like a large sphere, or would it be flat, as one more naturally, but quite possibly wrongly, imagines it?

A third possibility is that the universe is *negatively* curved. This means, not surprisingly, more or less the opposite of positively curved. Thus, evidence for negative curvature would be that the angles of a triangle added up to *less* than 180 degrees, that lines starting in the same direction tended to diverge, or that the circumference of a circle of radius r was *larger* than $2\pi r$. This sort of behaviour occurs in the hyperbolic disc. For example, Figure 37 shows a triangle whose angles add up to significantly less than 180 degrees. It is not hard to generalize the sphere and the hyperbolic

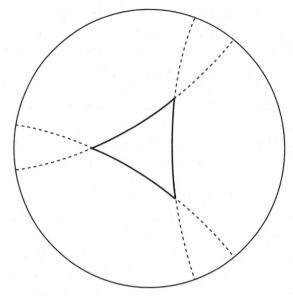

37. A hyperbolic triangle

disc to higher-dimensional analogues, and it could be that hyperbolic geometry is a better model for the large-scale shape of spacetime than either spherical or Euclidean geometry.

Manifolds

A closed surface means a two-dimensional shape that has no boundary. The surface of a sphere is a good example, as is a torus (the mathematical name for the shape of the surface of a quoit, or a ring-shaped doughnut). As the discussion of curvature made clear, it can be useful to think about such surfaces without reference to some three-dimensional space in which they live, and this becomes even more important if we want to generalize the notion of a closed surface to higher dimensions.

It is not just mathematicians who like to think about surfaces in a purely two-dimensional way. For example, the geometry of the United States is significantly affected by the curvature of the earth, but if one wishes to design a useful road map, it does not have to be printed on a single large curved piece of paper. Much more practical is to produce a book with several pages, each dealing with a small part of the country. It is best if these parts overlap, so that if a town lies inconveniently near the edge of one page, there will be another page where it doesn't. Moreover, at the edges of each page will be an indication of which other pages represent overlapping regions and how the overlap works. Because of the curvature of the earth, none of the pages will be exactly accurate, but one can include lines of constant latitude and longitude to indicate the small distortion, and in that way the geometry of the United States can be encapsulated in a book of flat pages.

There is nothing in principle to stop one producing an atlas that covers the whole world in similar detail (though many pages would be almost entirely blue). Therefore, the mathematical properties of a sphere can in a way be encapsulated in an atlas. If you want to answer geometrical questions about the sphere, are completely unable to visualize it, but have an atlas handy, then, with a bit of effort, you will be able to do it. Figure 38 shows a nine-page atlas, not of a sphere but of a torus. To see how it corresponds to a doughnut shape, imagine sticking the pages together to make one large page, then joining the top and bottom of the large page to form a cylinder, and finally bringing the two ends of the cylinder together and joining them.

One of the most important branches of mathematics is the study of objects known as manifolds, which result from generalizing these ideas to three or more dimensions. Roughly speaking, a d-dimensional manifold is any geometrical object in which every point is surrounded by a region that closely resembles a small piece of d-dimensional space. Since manifolds become much harder to visualize as the number of dimensions increases, the idea of an atlas becomes correspondingly more useful.

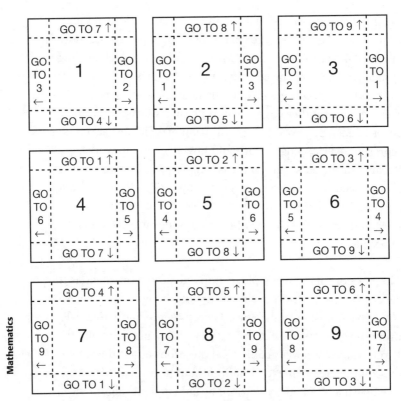

38. An atlas of a torus

Let us think for a moment what an atlas of a three-dimensional manifold would be like. The pages would of course have to be three-dimensional, and like the pages of a road map they would be flat. By that I mean that they would be chunks of familiar Euclidean space; one could require them to be cuboids, but this is not very important mathematically. Each of these three-dimensional 'pages' would be a map of a small part of the manifold, and one would carefully specify how the pages overlapped. A typical specification might be something like that the point (x, y, z) towards a particular edge of page A corresponded to the point $(2y, x, z-4)$ on page B.

Given such an atlas, how could one imagine moving about in the manifold? The obvious way would be to think of a point moving in one of the pages. If this point ever reached the edge of the page, there would be another page on which the same part of the manifold was represented, but where the point was not right at the edge, so one could turn to that page instead. Thus, the entire geometry of a manifold can be formulated in terms of an atlas, so that it is not necessary to think of the manifold as 'really' being a three-dimensional surface lying in a four-dimensional space. In fact, some three-dimensional manifolds cannot even be made to fit into four dimensions.

This idea of an atlas raises some natural questions. For example, although it enables us to say what happens if we move around in the manifold, how do we get from that information, which may be contained in a large number of pages with very complicated rules for how they overlap, to some feeling for the basic 'shape' of the manifold? How can we tell when two different atlases are actually atlases of the same manifold? In particular, is there some easy way of telling, by looking at a three-dimensional atlas, whether the manifold it represents is the three-dimensional surface of a four-dimensional sphere? A precise formulation of this last question, known as the Poincaré conjecture, is an open problem, for the solution of which a reward of one million dollars has been offered (by the Clay Mathematics Institute).

Chapter 7
Estimates and approximations

Most people think of mathematics as a very clean, exact subject. One learns at school to expect that if a mathematical problem can be stated succinctly, then it will probably have a short answer, often given by a simple formula. Those who continue with mathematics at university level, and particularly those who do research in the subject, soon discover that nothing could be further from the truth. For many problems it would be miraculous and totally unexpected if somebody were to find a precise formula for the solution; most of the time one must settle for a rough estimate instead. Until one is used to estimates, they seem ugly and unsatisfying. However, it is worth acquiring a taste for them, because not to do so is to miss out on many of the greatest theorems and most interesting unsolved problems in mathematics.

A simple sequence not given by a simple formula

Let $a_1, a_2, a_3, \ldots$ be real numbers generated by the following rule. The first number, a_1, equals 1, and thereafter each number equals the previous number plus its square root. In other words, for every n we let $a_{n+1} = a_n + \sqrt{a_n}$. This simply stated rule raises an obvious question: what is the number a_n?

To get a feel for the question, let us work out a_n for a few small values of n. We have $a_2 = 1 + \sqrt{1} = 1 + 1 = 2$. Then

$a_3 = a_2 + \sqrt{a_2} = 2 + \sqrt{2}$
$a_4 = a_3 + \sqrt{a_3} = 2 + \sqrt{2} + \sqrt{2 + \sqrt{2}}$
$a_5 = a_4 + \sqrt{a_4} = 2 + \sqrt{2} + \sqrt{2 + \sqrt{2}} + \sqrt{2 + \sqrt{2} + \sqrt{2 + \sqrt{2}}}$

and so on. Notice that the expressions on the right-hand side do not seem to simplify, and that each new one is twice as long as the previous one. From this observation it follows quite easily that the expression for a_{12} would involve 1,024 occurrences of the number 2, most of them deep inside a jungle of square-root signs. Such an expression would not give us much insight into the number a_{12}.

Should we therefore abandon any attempt to understand the sequence? No, because although there does not appear to be a good way of thinking about the *exact* value of a_n, except when n is very small, that does not rule out the possibility of obtaining a good estimate. Indeed, a good estimate may in the end be more useful. Above, I have written an exactly correct expression for a_5, but does that make you understand a_5 better than the information that a_5 is about seven and a half?

Let us therefore stop asking what a_n is and instead ask roughly how big a_n is. That is, let us try to find a simple formula that gives a good approximation to a_n. It turns out that such a formula exists: a_n is roughly $n^2/4$. It is a little tricky to prove this rigorously, but to see why it is plausible, notice that

$$(n + 1)^2/4 = (n^2 + 2n + 1)/4 = n^2/4 + n/2 + 1/4 = n^2/4 + \sqrt{n^2/4} + 1/4.$$

That is, if $b_n = n^2/4$, then $b_{n+1} = b_n + \sqrt{b_n} + 1/4$. If it were not for the '+ 1/4', this would tell us that the numbers b_n were generated exactly as the a_n are. However, when n is large, the addition of 1/4 is 'just a small perturbation' (this is the part of the proof that I am leaving out), so the b_n are generated *approximately* correctly, from which it can be deduced that b_n, that is, $n^2/4$, gives a good approximation to a_n, as I claimed.

Ways of approximating

It is important to specify what counts as a good approximation when making an assertion like this, because standards vary from context to context. If one is trying to approximate a sequence of steadily increasing numbers like $a_1, a_2, a_3, \ldots$ by a more easily defined sequence $b_1, b_2, b_3, \ldots$, then the best sort of approximation one can hope for, which is very rarely achieved, is one for which the difference between a_n and b_n is always below some fixed number – such as 1,000, for example. Then, as a_n and b_n become large, their *ratio* becomes very close to 1. Suppose, for example, that at some point $a_n = 2408597348632498758828$ and $b_n = 2408597348632498759734$. Then $b_n - a_n = 906$, which, though quite a large number, is tiny by comparison with a_n and b_n. If b_n approximates a_n in this sense, one says that a_n and b_n are 'equal up to an additive constant'.

Another good sort of approximation is one for which the ratio of a_n and b_n becomes very close to 1 as n gets large. This is true when a_n and b_n are equal up to an additive constant, but it is also true in other circumstances. For example, if $a_n = n^2$ and $b_n = n^2 + 3n$ then the ratio b_n/a_n is $1 + 3/n$, which is close to 1 for large n, even though the *difference* between a_n and b_n is $3n$, which is large.

Often even this is far too much to hope for, and one is happy to settle for still weaker notions of approximation. A common one is to regard a_n and b_n as approximately equal if they are 'equal up to a *multiplicative* constant'. This means that neither a_n/b_n nor b_n/a_n ever exceeds some fixed number – again, something like 1,000 would be a possibility, though the smaller the better. In other words, now it is not the difference between a_n and b_n that is kept within certain limits, but their ratio.

It may seem perverse to regard one number as being roughly the same as another number that is 1,000 times larger. But that is because we are used to dealing with small numbers. Of course, nobody would regard 17 as being roughly the same as 13,895, but it is not quite as ridiculous to say that two numbers like

2904756294089761562389545345987608890796872347514348757775468

and

3609823459872097612349870982498634087623457796784587345987166464

are, broadly speaking, of the same *sort* of size. Though the second is over 1,000 times larger, they both have about the same number of digits – between 60 and 65. In the absence of other interesting properties, that may well be all we care about.

When even *this* degree of approximation is asking too much, it is often still worthwhile to try to find two sequences $b_1, b_2, b_3, \ldots$ and $c_1, c_2, c_3, \ldots$ for which one can prove that b_n is always less than a_n, and c_n always greater. Then one says that b_n is a 'lower bound' for a_n, and c_n is an 'upper bound'. For example, a mathematician trying to estimate some quantity a_n might say, 'I don't know, even approximately, what a_n is, but I can prove that it is at least $n^2/2$ and no bigger than n^3.' If the problem is difficult enough, a theorem like this can be a significant achievement.

All you need to know about logarithms, square roots etc.

Part of the reason that the estimates and approximations that pervade mathematics are not well known outside the discipline is that in order to talk about them one uses phrases like 'about as fast as log n' or 'the square root of t, to within a constant', which mean little to most people. Fortunately, if one is concerned only with the *approximate* values of logarithms or square roots of large numbers, then they can be understood very easily, and so can this sort of language.

If you have two large positive integers m and n and you want a rough and ready estimate of their product mn, then what should you do? A good start is to count the digits of m and the digits of n. If m has h digits and n has k digits, then m lies between 10^{h-1} and 10^h, and n lies between 10^{k-1} and 10^k, which means that mn lies between 10^{h+k-2} and 10^{h+k}. Thus, merely by counting the digits of m and n, you can determine mn 'to within a factor of 100' – meaning that mn must lie between two numbers, 10^{h+k-2} and 10^{h+k}, and 10^{h+k} is only 100 times bigger than 10^{h+k-2}. If you compromise and go for 10^{h+k-1} as your estimate, then it will differ from mn by a factor of at most 10.

In other words, if you are interested in numbers only 'up to a multiplicative constant', then multiplication suddenly becomes very easy: take m and n, count their combined digits, subtract one (if you can be bothered), and write down a number with that many digits. For example, 1293875 (7 digits) times 20986759777 (11 digits) is in the region of 10000000000000000 (17 digits). If you want to be a little more careful, you can note that the first number begins with a 1 and the second with a 2, which means that 20000000000000000 is a better estimate, but for many purposes such precision is unnecessary.

Since approximate multiplication is easy, so is approximate squaring – just replace your number by a new one with twice as many digits. From this it follows that *halving* the number of digits of n approximates the square root of n. Similarly, dividing the number of digits by 3 approximates the cube root. More generally, if n is a large integer and t is any positive number, then n^t will have about t times as many digits as n.

What about logarithms? From the approximate point of view they are very simple indeed: the logarithm of a number is roughly the number of digits it has. For example, the logarithms of 34587 and 492348797548735 are about 5 and 15 respectively.

Actually, counting the digits of a number approximates its base-10 logarithm, the number you get by pressing LOG on a pocket calculator. Normally, when mathematicians talk about logarithms, they mean so-called 'natural' logarithms, which are logarithms to base e. Though the number e is indeed very natural and important, all we need to know here is that the natural logarithm of a number, the number you get by pressing LN on a calculator, is roughly the number of digits it has, multiplied by about 2.3. Thus, the natural logarithm of 2305799985748 is about $13 \times 2.3 = 29.9$. (If you know about logarithms, you will see that what you should really multiply by is $\log_e 10$.)

This process can be reversed. Suppose you have a number t and you know that it is the natural logarithm of another number n. This number n is called the *exponential* of t and is written e^t. What will n be, roughly? Well, to get t from n we count the number of digits of n and multiply by 2.3. Hence, the number of digits of n must be about $t/2.3$. This determines n, at least approximately.

The main use of the approximate definitions I have just given is that they enable one to make comparisons. For example, it is now clear that the logarithm of a large number n will be much smaller than its cube root: if n has 75 digits, for example, then its cube root will be very large – it has about 25 digits – but its natural logarithm will be only about $75 \times 2.3 = 172.5$. Similarly, the exponential of a number m will be much bigger than a power such as m^{10}: for example, if m has 50 digits, then m^{10} has around 500 digits, but the number of digits of e^m is about $m/2.3$, which is far bigger than 500.

The following table shows the approximate results of applying various operations to the number $n = 941192$. I have not included e^n, because if I had then I would have been obliged to change the title of this book.

n	941192
n^2	885842380864
$\sqrt{n}$	970.15
$\sqrt[3]{n}$	98
$\log_e n$	13.755
$\log_{10} n$	5.974

The prime number theorem

A prime number is a whole number greater than 1 that is divisible by no other whole numbers, with the obvious exceptions of 1 and itself. The prime numbers less than 150 are 2, 3, 5, 7, 11, 13, 17, 19, 23, 29, 31, 37, 41, 43, 47, 53, 59, 61, 67, 71, 73, 79, 83, 89, 97, 101, 103, 107, 109, 113, 127, 131, 137, 139, and 149. All other numbers less than 150 can be factorized: for example, $91 = 7 \times 13$. (You may be worried about the seemingly arbitrary exclusion of 1 from the definition of a prime. This does not express some deep fact about numbers: it just happens to be a useful convention, adopted so that there is only one way of factorizing any given number into primes.)

The primes have tantalized mathematicians since the Greeks, because they appear to be somewhat randomly distributed but not completely so. Nobody has found a simple rule that tells you what the nth largest prime is (of course one could laboriously write out a list of the first n primes, but this hardly counts as a simple rule, and would be completely impractical if n was large), and it is most unlikely that there is one. On the other hand, an examination of even the first 35 primes reveals some interesting features. If you work out the differences between successive primes, then you obtain the following new list: 1, 2, 2, 4, 2, 4, 2, 4, 6, 2, 6, 4, 2, 4, 6, 6, 2, 6, 4, 2, 6, 4, 6, 8, 4, 2, 4, 2, 4, 14, 4, 6, 2, 10. (That is, $1 = 3 - 2$, $2 = 5 - 3$, $2 = 7 - 5$, $4 = 11 - 7$, and so on.) This list is still somewhat disorderly, but the numbers in it have a tendency, just about discernible, to get gradually larger. Obviously they do not increase steadily, but numbers such as 10 and 14 do not appear until quite late on, while the first few are all 4 or under.

If you were to write out the first thousand primes, then the tendency for the gaps between successive ones to get larger would become more obvious. In other words, large primes appear to be thinner on the ground than small ones. This is exactly what one would expect, because there are more ways for a large number to *fail* to be prime. For example, one might guess that the number 10,001 was prime, especially as it is not divisible by 2, 3, 5, 7, 11, 13, 17, or 19 – but in fact it equals 73×137.

No self-respecting mathematician will be content with the mere observation (not even properly proved) that large primes are rarer than small ones. He or she will want to know *how much* rarer they are. If you choose a number at random between 1,000,001 and 1,010,000, then what are the chances that it will be a prime? In other words, what is the 'density' of primes near 1,000,000? Is it fantastically small or only quite small?

The reason such questions rarely occur to people who have not been exposed to university-level mathematics is that they lack the language in which to formulate and think about them. However, if you have understood this chapter so far, you are in a position to appreciate one of the greatest achievements of mathematics: the prime number theorem. This states that the density of primes near a number n is around $1/\log_e n$ – that is, one divided by the natural logarithm of n.

Consider once again the chances that a random number between 1,000,001 and 1,010,000 is prime. The numbers in this interval are all roughly equal to a million. The prime number theorem says that the density will therefore be around 1 divided by the natural logarithm of a million. The logarithm to base 10 is 6 (in this case, counting the digits would give 7, but since we know the exact answer we may as well use it), so the natural logarithm is about 6×2.3, or 13.8. Therefore, about 1 in 14 numbers between 1,000,001 and 1,010,000 is a prime, which works out at a little over 7% of them. By contrast, the number of primes less than 100 is 24, or about a quarter of the total, which illustrates how the density drops as they get larger.

Given the sporadic, random-like quality of the primes, it is quite surprising how much can be proved about them. Interestingly, theorems about the primes are usually proved by *exploiting* this seeming randomness. For example, a famous theorem of Vinogradov, proved in 1937, states that every large enough odd number can be written as the sum of three prime numbers. I cannot explain in this book how he proved this, but what he did not do is find a *method* for expressing odd numbers as sums of three primes. Such an approach would be almost certain to fail, because of the difficulty of generating even the primes themselves. Instead, building on work of Hardy and Littlewood, he argued roughly as follows. If you were to choose a genuinely random sequence of numbers of about the same density as the primes, then some elementary probability theory shows that you would almost certainly be able to write all large enough numbers as the sum of three members of your sequence. In fact, you would be able to do it in many different ways. Because the primes are random-like (the hard part of the proof is to say what this means and then prove it rigorously), their behaviour is similar to that of the random sequence, so all large enough numbers are the sum of three primes, also in many different ways. Just to illustrate this phenomenon, here are all the ways of writing 35 as a sum of three primes:

$$35 = 2 + 2 + 31 = 3 + 3 + 29 = 3 + 13 + 19 = 5 + 7 + 23$$

$$= 5 + 11 + 19 = 5 + 13 + 17 = 7 + 11 + 17 = 11 + 11 + 13.$$

Much research on prime numbers has this sort of flavour. You first devise a probabilistic model for the primes – that is, you pretend to yourself that they have been selected according to some random procedure. Next, you work out what would be true if the primes really were generated randomly. That allows you to guess the answers to many questions. Finally, you try to show that the model is realistic enough for your guesses to be approximately correct. Notice that such an approach would be impossible if you were forced to give exact answers at every stage of the argument.

It is interesting that the probabilistic model is a model not of a physical phenomenon, but of another piece of mathematics. Although the prime numbers are rigidly determined, they somehow feel like experimental data. Once we regard them that way, it becomes tempting to devise simplified models that allow us to predict what the answers to certain probabilistic questions are likely to be. And such models have indeed sometimes led people to proofs valid for the primes themselves.

Although this style of argument has had some notable successes, it has left open many famous problems. For example, Goldbach's conjecture asserts that every even number greater than 4 is the sum of two odd primes. This conjecture appears to be much more difficult than the three-primes question answered by Vinogradov. There is also the twin-primes conjecture, which states that there are infinitely many pairs of primes separated by 2, such as 17 and 19, or 137 and 139. Another way of putting this is that if you write out the successive differences, as I did above, then the number 2 keeps appearing for ever (though more and more rarely).

Perhaps the most famous open problem in mathematics is the Riemann hypothesis. This has several equivalent formulations. One of them concerns the accuracy of the estimate given by the prime number theorem. As I have said, the prime number theorem tells you the approximate density of the primes near any given number. From this information one can calculate roughly how many prime numbers there are up to any given number n. But how rough is rough? If $p(n)$ is the true number of primes up to n and $q(n)$ is the estimate suggested by the prime number theorem, then the Riemann hypothesis asserts that the difference between $p(n)$ and $q(n)$ will be not much larger than $\sqrt{n}$. If that sort of accuracy could be proved to hold, then it would have many applications, but what is known to date is far weaker.

Sorting algorithms

Another area of mathematics that is full of rough estimates is theoretical computer science. If one is writing a computer program to perform a certain task, then it is a good idea to design it in such a way that it will run as quickly as possible. Theoretical computer scientists ask the question: what is the fastest one could possibly hope for?

It is almost always unrealistic to ask for an exact answer to this question, so one tries to prove statements like, 'The following algorithm runs in about n^2 steps when the input size is n.' From this one can conclude that a typical PC will be able to handle an input size (roughly speaking, how much information you want it to analyse) of 1,000 but not one of 1,000,000. Thus, such estimates have a practical importance.

One very useful task that computers can do is known as sorting – that is, putting a large number of objects in order according to a given criterion. To think about this, imagine that you wish to arrange a collection of objects (not necessarily inanimate – they might, for example, be candidates for a job) in order of preference. Suppose that you cannot assign a numerical value to the amount that you like any given object, but that, given any two objects, you can always decide which you prefer. Suppose also that your preferences are consistent, in the sense that you never prefer A to B, B to C, and C to A. If you do not wish to spend long on the task, then it makes sense to try to minimize the number of comparisons you make.

When the number of objects is very small, it is easy to work out how to do this. For example, if there are two objects, then you must make at least one comparison, and once you have made it you know what order they are in. If there are three objects A, B, and C, then one comparison will not be enough, but you must start with some comparison and it doesn't matter which. Suppose, for the sake of argument, that you compare A with B and prefer A. Now you must compare one of A and B with C. If you compare A with C and prefer C to A, then you know that your order of preference is C, A, B. However, if, as may happen, you find that you prefer A to C, then all you know about B and C is that you prefer A to either of them. Then a third comparison will be needed so that B and C can be put in order. Hence, three comparisons are always sufficient and sometimes necessary.

What happens with four objects A, B, C, and D? The analysis becomes more complicated. You may as well start by comparing A with B. But once you have done that, there are two genuinely different possibilities for the next comparison. Either you can compare one of A and B with C, or you can compare C with D, and it is not clear which is the better idea.

Suppose you compare B with C. If you are lucky, then you will now be able to put A, B, and C in order. Suppose that this order is A, B, C. It then remains to see where D fits in. The best thing to do first is to compare D with B. After that, all you have to do is compare D with A (if you preferred D to B) or with C (if you preferred B to D). This makes a total of four comparisons – two to put A, B, and C in order and two to find out where to put D.

We have not finished analysing the problem, because you may not have been lucky with A, B, and C. Perhaps all you know after the first two comparisons is that both A and C are preferable to B. Then you have another dilemma: is it better to compare A with C or to compare D with one of A, B, and C – and in the second case, should D be compared with B or with one of A and C? And once you have finished looking at those cases and subcases, you still have to see what would have happened if your second comparison had been between C and D.

The analysis becomes somewhat tedious, but it can be done. It shows that five comparisons are always enough, that sometimes you need that many, and that the second comparison should indeed be between C and D.

The trouble with this sort of argument is that the number of cases one must consider becomes very large very quickly. It would be out of the question to work out exactly how many comparisons are needed when you have, say, 100 objects – almost certainly this will never be known. (I well remember my shock when I first heard a mathematician declare that the exact value of a certain quantity would never be known. Now I am used to the fact that this is the rule rather than the exception. The quantity in question was the Ramsey number $R(5, 5)$, the smallest number n for which in any group of n people there must be five who all know each other or five who are all new to each other.) Instead, therefore, one tries to find upper and lower bounds. For this problem, an upper bound of c_n means a procedure for sorting n objects using no more than c_n comparisons, and a lower bound of b_n means a proof that, no matter how clever you are, b_n comparisons will sometimes be necessary. This is an example of a problem where the best-known upper and lower bounds are within a multiplicative factor of each other: it is known that, up to a multiplicative constant, the number of comparisons needed to sort n objects is $n \log n$.

One way to see why this is interesting is to try to devise a sorting procedure for yourself. An obvious method is to start by finding the object that comes top, set it aside, and then repeat. To find the best object, compare the first two, then compare the winner with the third, and the winner of that with the fourth, and so on. This way, it takes $n - 1$ comparisons to find the best, then $n - 2$ to find the next best, and so on, making a total of $(n - 1) + (n - 2) + (n - 3) + \ldots + 1$ comparisons in all, which works out at about $n^2/2$.

Natural though this method is, if you use it, then you end up comparing *every* pair of objects, so it is in fact as inefficient as possible (though it does have the advantage of being simple to program). When n is large, $n \log n$ is a very significant improvement on $n^2/2$, because $\log n$ is much smaller than $n/2$.

The following method, known as Quicksort, is not guaranteed to work any faster, but usually it works much faster. It is defined recursively (that is, in terms of itself) as follows. First choose any one of the objects, x, say, and arrange the others into two piles, the ones that are better than x and the ones that are worse. This needs $n - 1$ comparisons. All you need to do now is sort the two piles – which you do using Quicksort. That is, for each pile you choose one object and arrange the others into two further piles, and so on. Usually, unless you are unlucky, when you divide a pile into two further ones, they will be of roughly the same size. Then it can be shown that the number of comparisons you make will be roughly $n \log n$. In other words, generally this method works as well as you could possibly hope for, to within a multiplicative constant.

Chapter 8
Some frequently asked questions

1. Is it true that mathematicians are past it by the time they are 30?

This widely believed myth derives its appeal from a misconception about the nature of mathematical ability. People like to think of mathematicians as geniuses, and of genius itself as an utterly mysterious quality which a few are born with, and which nobody else has the slightest chance of acquiring.

The relationship between age and mathematical output varies widely from one person to another, and it is true that a few mathematicians do their best work in their 20s. The great majority, though, find that their knowledge and expertise develop steadily throughout their life, and that for many years this development more than compensates for any decline that there might be in 'raw' brain power – if that concept even makes sense. It is true that not many major breakthroughs are achieved by mathematicians over the age of 40, but this may well be for sociological reasons. By the age of 40, somebody who is capable of making such a breakthrough will probably already have become well known as a result of earlier work, and may therefore not have quite the hunger of a younger, less established mathematician. But there are many counter-examples to this, and some mathematicians continue, with enthusiasm undimmed, well past retirement.

In general, the popular view of a stereotypical mathematician – very clever perhaps, but also peculiar, badly dressed, asexual, semi-autistic – is not a flattering one. A few mathematicians do conform to the stereotype to some extent, but nothing would be more foolish than to think that if you do not, then you cannot be any good at mathematics. Indeed, all other things being equal, you may well be at an advantage. Only a very small proportion of mathematics students end up becoming research mathematicians. Most fall by the wayside at an earlier stage, for example by losing interest, not getting a PhD place, or doing a PhD but not getting a university job. It is my impression, and I am not alone in thinking this, that, among those who do survive the various culls, there is usually a smaller proportion of oddballs than in the initial student population.

While the negative portrayal of mathematicians may be damaging, by putting off people who would otherwise enjoy the subject and be good at it, the damage done by the word 'genius' is more insidious and possibly greater. Here is a rough and ready definition of a genius: somebody who can do easily, and at a young age, something that almost nobody else can do except after years of practice, if at all. The achievements of geniuses have a sort of magic quality about them – it is as if their brains work not just more efficiently than ours, but in a completely different way. Every year or two a mathematics undergraduate arrives at Cambridge who regularly manages to solve in a few minutes problems that take most people, including those who are supposed to be teaching them, several hours or more. When faced with such a person, all one can do is stand back and admire.

And yet, these extraordinary people are not always the most successful research mathematicians. If you want to solve a problem that other professional mathematicians have tried and failed to solve before you, then, of the many qualities you will need, genius as I have defined it is neither necessary nor sufficient. To illustrate with an extreme example, Andrew Wiles, who (at the age of just over 40) proved Fermat's Last Theorem (which states that if x, y, z, and n are all positive integers and n is greater than 2, then $x^n + y^n$ cannot equal z^n) and thereby solved the world's most famous unsolved mathematics problem, is undoubtedly very clever, but he is not a genius in my sense.

How, you might ask, could he possibly have done what he did without some sort of mysterious extra brainpower? The answer is that, remarkable though his achievement was, it is not so remarkable as to defy explanation. I do not know precisely what enabled him to succeed, but he would have needed great courage, determination, and patience, a wide knowledge of some very difficult work done by others, the good fortune to be in the right mathematical area at the right time, and an exceptional strategic ability.

This last quality is, ultimately, more important than freakish mental speed: the most profound contributions to mathematics are often made by tortoises rather than hares. As mathematicians develop, they learn various tricks of the trade, partly from the work of other mathematicians and partly as a result of many hours spent thinking about mathematics. What determines whether they can use their expertise to solve notorious problems is, in large measure, a matter of careful planning: attempting problems that are likely to be fruitful, knowing when to give up a line of thought (a difficult judgement to make), being able to sketch broad outlines of arguments before, just occasionally, managing to fill in the details. This demands a level of maturity which is by no means incompatible with genius but which does not always accompany it.

2. Why are there so few women mathematicians?

It is tempting to avoid this question, since giving an answer presents such a good opportunity to cause offence. However, the small proportion of women, even today, in the mathematics departments of the world, is so noticeable and so very much a fact of mathematical life, that I feel compelled to say something, even if what I say amounts to little more than that I find the situation puzzling and regrettable.

One point that deserves to be made is that the lack of women in mathematics is another statistical phenomenon: incredibly good female mathematicians exist and, just like their male counterparts, they have many different ways of being good, including, in some cases, being geniuses. There is no evidence whatsoever for any sort of upper limit on what women can achieve mathematically. Occasionally, one reads that men perform better at certain mental tests, of visuo-spatial ability, for example, and it is sometimes suggested that this accounts for their domination of mathematics. However, this sort of argument is not very convincing: visuo-spatial ability can be developed with practice, and, in any case, while it can sometimes be useful to a mathematician, it is rarely indispensable.

More plausible is the idea that social factors are important: where a boy may be proud of his mathematical ability, one can imagine a girl being embarrassed to excel at a pursuit that is perceived as unfeminine. In addition, mathematically gifted girls have few role models, so the situation is self-perpetuating. A social factor that may operate at a later stage is that mathematics, more than most academic disciplines, demands a certain single-mindedness which is hard, though certainly not impossible, to combine with motherhood. The novelist Candia McWilliam once said that each of her children had cost her two books; but at least it is possible to write a novel after a few years of not having done so. If you give up mathematics for a few years then you get out of the habit, and mathematical comebacks are rare.

It has been suggested that female mathematicians tend to develop later than their male counterparts and that this puts them at a disadvantage in a career structure that rewards early achievement. The life stories of many of the most prominent female mathematicians bear this out, though their late development is largely for the social reasons just mentioned, and again there are many exceptions.

None of these explanations seems sufficient, though. Rather than speculating any further, the best I can do is point out that several books have been written on this subject (see Further reading). A final comment is that the situation is improving: the proportion of women among mathematicians has steadily increased in recent years and, given how society in general has changed and is changing, it will almost certainly continue to do so.

3. Do mathematics and music go together?

Despite the fact that many mathematicians are completely unmusical and few musicians have any interest in mathematics, there is a persistent piece of folk knowledge that the two are connected. As a result, nobody is surprised to learn that a mathematician is a very good pianist, or composes music as a hobby, or loves to listen to Bach.

There is plenty of anecdotal evidence suggesting that mathematicians are drawn to music more than to any other art form, and some studies have claimed to demonstrate that children who are educated musically perform better in scientific subjects. It is not hard to guess why this might be. Although abstraction is important in all the arts, and music has a representational component, music is the most obviously abstract art: a large part of the pleasure of listening to music comes from a direct, if not wholly conscious, appreciation of pure patterns with no intrinsic meaning.

Unfortunately, the anecdotal evidence is backed up by very little hard science. It is not even obvious what questions should be asked. What would we learn if statistically significant data were collected showing that a higher percentage of mathematicians played the piano than of non-mathematicians with similar social and educational backgrounds? My guess is that such data *could* be collected, but it would be far more interesting to produce an experimentally testable theory that explained the connection. As for statistical evidence, this would be much more valuable if it was more specific. Both mathematics and music are very varied: it is possible to be passionately enthusiastic about some parts and completely uninterested in others. Are there subtle relationships between mathematical and musical tastes? If so, then they would be much more informative than crude correlations between interest in the disciplines as a whole.

4. Why do so many people positively dislike mathematics?

One does not often hear people saying that they have never liked biology, or English literature. To be sure, not everybody is excited by these subjects, but those who are not tend to understand perfectly well that others are. By contrast, mathematics, and subjects with a high mathematical content such as physics, seem to provoke not just indifference but actual antipathy. What is it that causes many people to give mathematical subjects up as soon as they possibly can and remember them with dread for the rest of their lives?

Probably it is not so much mathematics itself that people find unappealing as the experience of mathematics lessons, and this is easier to understand. Because mathematics continually builds on itself, it is important to keep up when learning it. For example, if you are not reasonably adept at multiplying two-digit numbers together, then you probably won't have a good intuitive feel for the distributive law (discussed in Chapter 2). Without this, you are unlikely to be comfortable with multiplying out the brackets in an expression such as $(x + 2)(x + 3)$, and then you will not be able to understand quadratic equations properly. And if you do not understand quadratic equations, then you will not understand why the golden ratio is $\dfrac{1 + \sqrt{5}}{2}$.

There are many chains of this kind, but there is more to keeping up with mathematics than just maintaining technical fluency. Every so often, a new idea is introduced which is very important and markedly more sophisticated than those that have come before, and each one provides an opportunity to fall behind. An obvious example is the use of letters to stand for numbers, which many find confusing but which is fundamental to all mathematics above a certain level. Other examples are negative numbers, complex numbers, trigonometry, raising to powers, logarithms, and the beginnings of calculus. Those who are not ready to make the necessary conceptual leap when they meet one of these ideas will feel insecure about all the mathematics that builds on it. Gradually they will get used to only half understanding what their mathematics teachers say, and after a few more missed leaps they will find that even half is an overestimate. Meanwhile, they will see others in their class who are keeping up with no difficulty at all. It is no wonder that mathematics lessons become, for many people, something of an ordeal.

Is this a necessary state of affairs? Are some people just doomed to dislike mathematics at school? Or might it be possible to teach the subject differently in such a way that far fewer people are excluded from it? I am convinced that any child who is given one-to-one tuition in mathematics from an early age by a good and enthusiastic teacher will grow up liking it. This, of course, does not immediately suggest a feasible educational policy, but it does at least indicate that there might be room for improvement in how mathematics is taught.

One recommendation follows from the ideas I have emphasized in this book. Above, I implicitly drew a contrast between being technically fluent and understanding difficult concepts, but it seems that almost everybody who is good at one is good at the other. And indeed, if understanding a mathematical object is largely a question of learning the rules it obeys rather than grasping its essence, then this is exactly what one would expect – the distinction between technical fluency and mathematical understanding is less clear-cut than one might imagine.

How should this observation influence classroom practice? I do not advocate any revolutionary change – mathematics has suffered from too many of them already – but a small change in emphasis could pay dividends. For example, suppose that a pupil makes the common mistake of thinking that $x^{a+b} = x^a + x^b$. A teacher who has emphasized the intrinsic meaning of expressions such as x^a will point out that x^{a+b} means $a + b$ xs all multiplied together, which is clearly the same as a of them multiplied together *multiplied* by b of them multiplied together. Unfortunately, many children find this argument too complicated to take in, and anyhow it ceases to be valid if a and b are not positive integers.

Such children might benefit from a more abstract approach. As I pointed out in Chapter 2, everything one needs to know about powers can be deduced from a few very simple rules, of which the most important is $x^{a+b} = x^a x^b$. If this rule has been emphasized, then not only is the above mistake less likely in the first place, but it is also easier to correct: those who make the mistake can simply be told that they have forgotten to apply the right rule. Of course, it is important to be familiar with basic facts such as that x^3 means x times x times x, but these can be presented as consequences of the rules rather than as justifications for them.

I do not wish to suggest that one should try to explain to children what the abstract approach is, but merely that teachers should be aware of its implications. The main one is that it is quite possible to learn to use mathematical concepts correctly without being able to say exactly what they mean. This might sound a bad idea, but the use is often easier to teach, and a deeper understanding of the meaning, if there *is* any meaning over and above the use, often follows of its own accord.

5. Do mathematicians use computers in their work?

The short answer is that most do not, or at least not in a fundamental way. Of course, just like anybody else, we find them indispensable for word-processing and for communicating with each other, and the Internet is becoming more and more useful. There are some areas of mathematics where long, unpleasant but basically routine calculations have to be done, and there are very good symbolic manipulation programs for doing them.

Thus, computers can be very useful time-saving devices, sometimes so much so that they enable mathematicians to discover results that they could not have discovered on their own. Nevertheless, the kind of help that computers can provide is very limited. If it happens that your problem, or more usually sub-problem, is one of the small minority that can be solved by a long and repetitive search, then well and good. If, on the other hand, you are stuck and need a bright idea, then, in the present state of technology, a computer will be no help whatsoever. In fact, most mathematicians would say that their most important tools are a piece of paper and something to write with.

My own view, which is a minority one, is that this is a temporary state of affairs, and that, over the next hundred years or so, computers will gradually be able to do more and more of what mathematicians do – starting, perhaps, with doing simple exercises for us or saving us from wasting a week trying to prove a lemma to which a well-known construction provides a counter-example (here I speak from frequent experience) and eventually supplanting us entirely. Most mathematicians are far more pessimistic (or should that be optimistic?) about how good computers will ever be at mathematics.

6. How is research in mathematics possible?

Conversely, one might ask, what is it that seems so paradoxical about the possibility of mathematical research? I have mentioned several unsolved problems in this book, and mathematical research consists, in large measure, in trying to solve those and similar ones. If you have read Chapter 7, then you will see that a good way to generate questions is to take a mathematical phenomenon that is too hard to analyse exactly, and try to make approximate statements about it. Another method is suggested by the end of Chapter 6: choose a difficult mathematical concept, such as a four-dimensional manifold, and you will usually find that even simple questions about it can be very hard to answer.

If there is a mystery about mathematical research, it is not that hard questions exist – it is in fact quite easy to invent impossibly hard questions – but rather that there are enough questions of just the right level of difficulty to keep thousands of mathematicians hooked. To do this, they must certainly be challenging but they must also offer a glimmer of hope that they can be solved.

7. Are famous mathematical problems ever solved by amateurs?

The simplest and least misleading answer to this question is a straightforward no. Professional mathematicians very soon learn that almost any idea they have about any well-known problem has been had by many people before them. For an idea to have a chance of being new, it must have some feature that explains why nobody has previously thought of it. It may be simply that the idea is strikingly original and unexpected, but this is very rare: on the whole, if an idea comes, it comes for a good reason rather than simply bubbling up out of nowhere. And if it has occurred to you, then why should it not have occurred to somebody else? A more plausible reason is that it is related to other ideas which are not particularly well known but which you have taken the trouble to learn and digest. That at least reduces the probability that others have had it before you, though not to zero.

Mathematics departments around the world regularly receive letters from people who claim to have solved famous problems, and virtually without exception these 'solutions' are not merely wrong, but laughably so. Some, while not exactly mistaken, are so unlike a correct proof of anything that they are not really attempted solutions at all. Those that follow at least some of the normal conventions of mathematical presentation use very elementary arguments that would, had they been correct, have been discovered centuries ago. The people who write these letters have no conception of how difficult mathematical research is, of the years of effort needed to develop enough knowledge and expertise to do significant original work, or of the extent to which mathematics is a collective activity.

By this last point I do not mean that mathematicians work in large groups, though many research papers have two or three authors. Rather, I mean that, as mathematics develops, new techniques are invented that become indispensable for answering certain kinds of questions. As a result, each generation of mathematicians stands on the shoulders of previous ones, solving problems that would once have been regarded as out of reach. If you try to work in isolation from the mathematical mainstream, then you will have to work out these techniques for yourself, and that puts you at a crippling disadvantage.

This is not quite to say that no amateur could ever do significant research in mathematics. Indeed, there are one or two examples. In 1975 Marjorie Rice, a San Diego housewife with very little mathematical training, discovered three previously unknown ways of tiling the plane with (irregular) pentagons after reading of the problem in the *Scientific American*. And in 1952 Kurt Heegner, a 58-year-old German schoolmaster, proved a famous conjecture of Gauss which had been open for over a century.

However, these examples do not contradict what I have been saying. There are some problems which do not seem to relate closely to the main body of mathematics, and for those it is not particularly helpful to know existing mathematical techniques. The problem of finding new pentagonal tilings was of such a kind: a professional mathematician would not have been much better equipped to solve it than a gifted amateur. Rice's achievement was rather like that of an amateur astronomer who discovers a new comet – the resulting fame is a well-deserved reward for a long search. As for Heegner, though he was not a professional mathematician, he certainly did not work in isolation. In particular, he had taught himself about modular functions. I cannot explain what these are here – indeed, they would normally be considered too advanced even for an undergraduate mathematics course.

Interestingly, Heegner did not write up his proof in a completely conventional way, and although his paper was grudgingly published it was thought for many years to be wrong. In the late 1960s, the problem was solved again, independently, by Alan Baker and Harold Stark, and only then was Heegner's work carefully re-examined and found to be correct after all. Unfortunately, Heegner died in 1965 and thus did not live to see his rehabilitation.

8. Why do mathematicians refer to some theorems and proofs as beautiful?

I have discussed this question earlier in the book, so here I will be very brief. It may seem odd to use aesthetic language about something as apparently dry as mathematics, but, as I explained on page 51 (at the end of the discussion of the tiling problem), mathematical arguments can give pleasure, and this pleasure has many features in common with more conventional aesthetic pleasure.

One difference is that, at least from the aesthetic point of view, a mathematician is more anonymous than an artist. While we may greatly admire a mathematician who discovers a beautiful proof, the human story behind the discovery eventually fades away and it is, in the end, the mathematics itself that delights us.

Further reading

There are some important aspects of mathematics that I have not had the space to discuss here. For these I can recommend other books. If you want to learn about the history of mathematics, it is hard to beat Morris Kline's magisterial three volumes on the subject, *Mathematical Thought from Ancient to Modern Times* (Oxford University Press, 1972), though he expects more mathematical sophistication from his readers than I have here. *Innumeracy*, by John Allen Paulos, (Viking, 1989) has rapidly become a classic on the subject of how knowledge of mathematics can influence one's judgements in everyday life—for the better. Tom Körner's *The Pleasures of Counting* (Cambridge University Press, 1996) says much more about the applications of mathematics than I have, and does so more wittily. *What is Mathematics?* by Courant and Robbins (Oxford University Press, 2nd edn., 1996) is another classic. It is similar in spirit to this book, but longer and somewhat more formal. *The Mathematical Experience* by Davis and Hersch (Birkhäuser, 1980) is a delightful collection of essays about mathematics, written in a philosophical vein. I would have liked to say more about probability, but a beautiful discussion of randomness and its philosophical implications can instead be found in *Mathematics and the Unexpected*, by Ivar Ekeland (University of Chicago Press, 1988).

The quotations on page 18 are from Saussure's *Course in General Linguistics* (McGraw-Hill, 1959) and Wittgenstein's *Philosophical Investigations* (Blackwell, 3rd edn., 2001). Anybody who has read this book and the *Philosophical Investigations* will see how much the later Wittgenstein has influenced my philosophical outlook and in particular my views on the abstract method. Russell and Whitehead's famous *Principia Mathematica* (Cambridge University Press, 2nd edn., 1973) is not exactly light reading, but if you found some of my proofs of elementary facts long-winded, then for comparison you should look up their proof that $1 + 1 = 2$. On the subject of women in mathematics, discussed in Chapter 8, two good recent books are *Women in Mathematics: The Addition of Difference* by Claudia Henrion (Indiana University Press, 1997) and *Women Becoming Mathematicians: Creating a Professional Identity in Post-World War II America* by Margaret Murray (MIT Press, 2000).

Finally, if you have enjoyed this book, you might like to know that in order to keep it very short I reluctantly removed whole sections, including a complete chapter, from earlier drafts. Some of this material can be found on my home page:

http://www.dpmms.cam.ac.uk/~wtg10

"牛津通识读本"已出书目

古典哲学的趣味	福柯	地球
人生的意义	缤纷的语言学	记忆
文学理论入门	达达和超现实主义	法律
大众经济学	佛学概论	中国文学
历史之源	维特根斯坦与哲学	托克维尔
设计，无处不在	科学哲学	休谟
生活中的心理学	印度哲学祛魅	分子
政治的历史与边界	克尔凯郭尔	法国大革命
哲学的思与惑	科学革命	民族主义
资本主义	广告	科幻作品
美国总统制	数学	罗素
海德格尔	叔本华	美国政党与选举
我们时代的伦理学	笛卡尔	美国最高法院
卡夫卡是谁	基督教神学	纪录片
考古学的过去与未来	犹太人与犹太教	大萧条与罗斯福新政
天文学简史	现代日本	领导力
社会学的意识	罗兰·巴特	无神论
康德	马基雅维里	罗马共和国
尼采	全球经济史	美国国会
亚里士多德的世界	进化	民主
西方艺术新论	性存在	英格兰文学
全球化面面观	量子理论	现代主义
简明逻辑学	牛顿新传	网络
法哲学：价值与事实	国际移民	自闭症
政治哲学与幸福根基	哈贝马斯	德里达
选择理论	医学伦理	浪漫主义
后殖民主义与世界格局	黑格尔	批判理论

德国文学	儿童心理学	电影
戏剧	时装	俄罗斯文学
腐败	现代拉丁美洲文学	古典文学
医事法	卢梭	大数据
癌症	隐私	洛克
植物	电影音乐	幸福
法语文学	抑郁症	免疫系统
微观经济学	传染病	银行学
湖泊	希腊化时代	景观设计学
拜占庭	知识	神圣罗马帝国
司法心理学	环境伦理学	大流行病
发展	美国革命	亚历山大大帝
农业	元素周期表	气候
特洛伊战争	人口学	第二次世界大战
巴比伦尼亚	社会心理学	中世纪
河流	动物	工业革命
战争与技术	项目管理	传记
品牌学	美学	公共管理
数学简史	管理学	社会语言学
物理学	卫星	物质
行为经济学	国际法	学习
计算机科学	计算机	化学
儒家思想		